CORNELL STUDIES IN CIVIL LIBERTY

ROBERT E. CUSHMAN, ADVISORY EDITOR

SECURITY, LOYALTY, AND SCIENCE

D1028622

Security, Loyalty, and Science

★

WALTER GELLHORN

PROFESSOR OF LAW IN COLUMBIA UNIVERSITY

Cornell University Press

ITHACA, NEW YORK, 1950

Open access edition funded by the National Endowment for the
Humanities/Andrew W. Mellon Foundation Humanities Open Book
Program.

First paperback printing 2019

ISBN 978-1-5017-4067-1 (pbk.: alk. paper)
ISBN 978-1-5017-4068-8 (pdf)
ISBN 978-1-5017-4069-5 (epub/mobi)

Librarians: A CIP catalog record for this book is available from the
Library of Congress

Preface

THIS volume is one of a series made possible by a grant from the Rockefeller Foundation to Cornell University. For two years a group of scholars working individually under my direction have studied the impact upon our civil liberties of current governmental programs designed to ensure internal security and to expose and control disloyal or subversive conduct. This research has covered federal and state legislative activities in this area, the operation of federal and local loyalty programs, and this book by Professor Walter Gellhorn of the Columbia University School of Law is a study of the administration of security policies in "sensitive" areas. Other volumes in the series include one on the House Committee on Un-American Activities, by Professor Robert K. Carr of Dartmouth College; one on the President's loyalty program and the summary dismissal statutes, by Miss Eleanor Bontecou, formerly an attorney in the Department of Justice; and a survey of state programs for the control of subversive activities, by several scholars working under Professor Gellhorn's editorship. There are monographs dealing with California, by Edward L. Barrett, Jr., of the University of California School of Law; with New York, by Lawrence H. Chamberlain, Dean of Columbia College; and with Washington, by Vern Countryman of the Yale Law School. A final report summarizes the findings of the entire study.

No thoughtful person will deny or minimize the need for protecting, and protecting adequately, our national security. The right and duty of national self-preservation cannot be challenged. This protection of the national security requires in certain instances the restriction of some of our traditional civil liberties. We have, however, learned by hard experience that we can be made to sacrifice more civil liberty to the cause of national security than is really necessary. There is, therefore, sound reason for examining with objective care the appropriateness and effectiveness of any particular governmental action sought to be justified as a defensive measure against disloyal or subversive persons or conduct. This is what the books in this series undertake to do, and Professor Gellhorn's present study deals with an area in which our national security exacts perhaps its heaviest toll in terms of the normal individual freedoms which must be restricted.

It must be emphasized that the volumes in this series state the views, conclusions, and recommendations of the individual authors. An advisory committee of distinguished men has been associated with this project. They are Messrs. Lloyd K. Garrison of New York, Erwin N. Griswold of Cambridge, Earl G. Harrison of Philadelphia, and Philip L. Graham of Washington. Each volume in the series has been strengthened and improved by the advice and suggestions of this committee, but each volume still remains the work and states the opinions of the person who wrote it.

ROBERT E. CUSHMAN

Cornell University
Ithaca, New York

Contents

SECURITY, LOYALTY, AND SCIENCE

Introduction

THE world's polarization into opposing forces has cast a shadow upon the traditionally accepted values of scientists. In days gone by science was broadly viewed as an unselfish effort, international in scope, to expand knowledge for the benefit of all mankind. Today science has come to be regarded somewhat in the nature of a national war plant in which a fortune has been invested.

The ties between government and science in the United States are increasingly tight. The Federal Government alone expends more than a billion dollars annually to support well over 50 percent of all the country's scientific research endeavors. In part this support is untinctured by the martial flavor of the times. Studies looking toward preservation of health or conservation of natural resources, toward agricultural abundance or aviation safety, would go forward with equal, perhaps even greater, intensity if peace were in the air. But since the atmosphere is not wholly restful, the prevailing emphasis is on studies related somehow to war. Few major industrial or institutional laboratories are without Army, Navy, Air Force, or Atomic Energy Commission contracts. Military research and development contracts alone number close to 20,000, at a cost each year in the neighborhood of $600,000,000. This means that nearly four cents of every dollar appropriated for the use of the armed forces, or about one cent of every dollar paid in federal taxes, is spent for research looking toward more effective weapons, equipment, medicines, and utilization of human resources in war. To this must still be added the research monies disbursed by the Atomic Energy Commission and many other civilian agencies as part of their respective programs.

These massive expenditures are acknowledgments of the immense contributions of science toward winning the most recent war—radar, the proximity fuze, the atomic bomb, the lifesaving drugs, and all the smaller mechanisms and techniques that were woven into the normality of military operations. They reflect, too, an awareness that the perils of the future may include still further extensions of military science. The average citizen, it is fair to suppose, is well persuaded that the remote and mysterious laboratory is the very citadel of his defense and the outpost whence to launch attack if need be.

So it is that the old picture of science as the universal benefactor has become somewhat eclipsed by a less lovely picture of science as an armory of devices for waging war more efficiently than any enemy.

Possession of this armory by the United States has not proved to be a wholly unmixed delight. This nation's comfortable consciousness of power is modified by anxious concern lest the armory be invaded by others who themselves seek the knowledge and instruments that constitute military superiority.

To prevent this, physical safeguards are erected. Fences and guards exclude unauthorized persons from scientific laboratories as from ordinary war plants. An Army ground division as well as Air Force units figures in the protection of the Atomic Energy Commission's installation at Hanford in Washington. Special squads of FBI agents are given technical

2

indoctrination courses and are then stationed in AEC laboratories. The Los Alamos area is patrolled by uniformed troopers of the Security Service, who far outnumber the scientists in the quarters under guard. Studies of sabotage vulnerability are made and protective measures are initiated at each of the more than 1,300 locations in the United States where work is done in connection with the atomic energy project alone. In addition to military and FBI personnel, some seven thousand persons whose salaries are paid by the Atomic Energy Commission devote full time to guard details and other aspects of "security."

These protections, however, are not enough, for the analogy between the laboratory and the ordinary war plant is incomplete. In science as it relates to military advantage, the great fear is that a competitor foreign nation, specifically the Soviet Union, may learn what American scientists have discovered and may thus diminish this country's margin of real or supposed superiority. Physical barriers may prevent access to areas where work is being done, but they do not furnish full assurance that ideas and information will not pass beyond the enclosed areas. The desired safety must be achieved, if at all, by other devices. This book is about those devices and their consequences.

The first thing to be noted is that, in the name of security, the United States has restricted the interchange of ideas between one scientist and another. How this has been done, how information has become "classified" (in the parlance of the military authorities) or "restricted" (in the parlance of the Atomic Energy Commission), furnishes the material of the opening chapter.

Obviously, however, it is not enough to say simply that the United States thinks it possesses secrets which it desires to withhold from others. Distinguished scientists advised from the first that scientific knowledge could not be monopolized

3

and that even the closely guarded "secret of the atomic bomb" would not long remain ours alone. The disclosure in the autumn of 1949 that there had been an atomic explosion in the Soviet Union served to demonstrate the soundness of this advice in point of fact, but the question remained whether a mere retardation of scientific work in other countries might not in itself be advantageous to this one. That question is considered in Chapter II, "The Balance Sheet of Secrecy." Whatever be the gains from suppressing the normal flow of scientific data, the costs also must be weighed before the validity of the policy may be assessed finally.

It is arguable that the United States is purchasing *security* at the price of *progress*. A secrecy program is marked mainly by apprehensive and backward glances over one's shoulder, and this may, in short, retard the forward drive of scientific energies into as yet unexplored areas. This phase of the problem warrants close and dispassionate attention. Critics of the present rigidities of secrecy policy have too often been dismissed as impractical sentimentalists or as plainly pro-Russian. Grave matters are involved. They should be considered with realistic detachment rather than with the preconceived notion that truth, if disagreeably comfortless, is unpatriotic. David Lilienthal in one of his last speeches as chairman of the Atomic Energy Commission declared that "we should stop this senseless business of choking ourselves by some of the extremes of secrecy to which we have been driven, extremes of secrecy that impede our own technical progress and our own defense." It would be reckless to ignore the facts one learns from so authoritative a source.

Secrecy is not the only step by which the goal of national safety is sought. The United States, like other countries, has placed selective limitations upon the persons who may engage in some types of scientific work. To some extent this is a direct reinforcement of secrecy regulations, being but a

4

means of identifying and accrediting the persons to whom secrets may be communicated. In part, however, an independent consideration enters into personnel restrictions. The position of scientists in contemporary society has been sharply affected by collective fear of Communist influences at home and abroad as threats to American security and independence. The Communists and their more or less formal allies have a scant record of accomplishment or influence in this country. But they are linked ideologically and emotionally to the Soviet Union, the only nation remotely capable of forcefully challenging the military dominance of the United States. Hence they are generally the object of the distrust and disquietude which reflect America's tensions. Since the dread of war underlies many other anxieties, and since the ingenuity of modern science and engineering serves constantly to intensify that dread, it is but natural that the scientist is an especial focus of the pervasive concern about Communists. In later chapters the "security" and "loyalty" programs are discussed in relation to scientists and their work; these are the programs that largely determine who can undertake what researches in America, and where and how.

As in the case of secrecy, an appraisal of the worth of these programs cannot be made solely in the light of their possible advantages. They entail costs, too. It may be that the nation loses more than it gains when, in order to pass on a scientist's eligibility to participate in research, it seeks to examine and confine his political attitudes, his personal associations, and his intellectual drifts. In any event, that question can best be considered after a description of the applicable policies and their administration.

The final answer will not be found in legal propositions, or in constitutional judgments. The Constitution in some circumstances sets a standard of propriety, to be sure; but it is never more than a minimum standard. Much that may be permissible may not be desirable. In this volume little effort has been made to spell out arguments about the legality or illegality of the courses the nation is following in its treatment of scientific personnel. The issues at stake are deeper than those with which courts customarily deal. If what is being done is in truth desirable, no doubt the appropriate supports can be discovered in law. If what is being done is in truth a disservice to the nation, it must be revised whether or not it is objectionable in a lawyer's sense.

A civilized nation, it has been remarked, is one that cannot tolerate wrongs or injustices—except at home. Even if this salty comment were unqualifiedly exact, the United States could not ignore the importance of finding out whether the tests applied to scientists create injuries without fully compensatory advantages. For it is clearly true, as President Truman told the American Association for the Advancement of Science on September 13, 1948, "We cannot drive scientists into our laboratories, but, if we tolerate reckless or unfair attacks, we can certainly drive them out." The following chapters about the measures which this country has adopted for purposes of self-protection seek to discover whether they serve as an adequate shield against enemies or, instead, as an unintended slashing of the human values that are the strongest elements of the American fabric.

It is not only modern warfare that rests upon technological achievement. Modern civilization does so as well. The preservation and advancement of society will be heavily affected, if not altogether determined, by the tone and quality of future scientific researches. In the United States the relationship between the nation's government and the nation's science is likely to grow closer rather than more distant, because it seems probable that only the Government can readily bear the burden of supporting research that is not immediately productive of profit. While ultimately the organizational

6

forms may change, with direction passing from military to civilian hands and with renewed emphasis upon scientific contributions to life rather than to death, the behavior patterns of today will help shape tomorrow. Present security methods and attitudes bear upon scientific advance. That is why they must be explored, identified, and understood.

A further word needs to be said about espionage in this era of international friction. Many persons of wide experience and cool judgment regard our present position vis-à-vis the Soviet Union as perilous in the extreme. In a situation which borders on national emergency, security measures become not only palatable but essential. Moreover, the case of Klaus Fuchs, the British atomic scientist who confessed to a long course of betrayal, has underscored the fact that treachery is more than a theoretical possibility.

Fuchs was an outstanding and trusted scientific worker. His self-exposure as a spy produced an altogether understandable shock of alarm. Fuchs's unmasking is a salutary reminder that in any large group of highly placed men, there may be some who are corrupt or cowardly or hostile. Whether those men are scientists or not, their detection and separation from positions of responsibility is of course a matter of importance.

Some nonscientists smugly suppose that but for Fuchs's revelation of secrets, the Russians would have been incapable of constructing an atomic bomb. They like to feel that American technology is so superior that other countries will remain baffled by scientific problems we have solved, unless the others succeed in stealing our solutions. If this view prevails, one can anticipate an intensified isolation of American science, an even sterner restraint upon discussion of researches, and a sharply suspicious attitude toward the individuals who perforce know about American scientific developments.

But the lesson of the Fuchs case will have been utterly missed if we blindly accept ever more rigid controls in the hope that security will thus, and only thus, be won. The Russians' achievement of a bomb may indeed have been materially advanced by Fuchs's messages. Responsible scientists, however, are agreed that espionage (even by one so well-informed as was Fuchs) could have had no effectiveness whatsoever unless the Soviet Union were already capable of exploiting the known facts. In the editorial words of the *Bulletin of the Atomic Scientists*, "No spying could have enabled a scientifically and industrially backward state to produce an atomic bomb in five, six, or twenty years." Fuchs's dereliction of duty was grave. So, too, would be the misdeeds of other spies who may conceivably have found employment in American scientific establishments. Grave as they could perhaps be, these misdeeds might still cost the United States less dearly than would excessively rigorous controls. As the following chapters suggest, there are dangers in damming, as well as dangers in wholly unblocking, the streams of knowledge. There are dangers, too, in overcautious selection of the scientists in whom trust is to be placed. American strength rests upon advance rather than upon nervous hoarding of present scientific knowledge. If Fuchs's treachery leads the American public to overlook that fact, this country will indeed have paid heavily for his faithlessness.

I

Keeping Secrets

EVEN before the United States became a participant in
World War II, many American scientists had customarily
worked in the atmosphere of suspicion engendered by secrecy.
So there is nothing entirely novel about censorship and security controls in research centers. Not until 1945, however, did
the dramatic detonations of the atomic bomb bring to general attention the extent to which major endeavors could be
carried on without public awareness.

Partly because they themselves were successfully kept from
knowing about the bomb until it had burst, many Americans
have considerable faith in the feasibility of keeping secrets.
This faith has not on the whole been a product of full reflection as to the possible undesirability of secrecy, or of awareness that secretiveness may not be practical in all circumstances.

At the present time the security policies of the United
States look toward the preservation of two distinct types of
secret. One of these is exemplified by the number of atomic
bombs which have been produced, or their whereabouts. If information concerning these matters is not volunteered, stolen,
or extorted, they will remain true secrets, not discoverable by
research because they are not facts in nature.

The other type of "secret" is exemplified by the exact number of neutrons created in the fission of plutonium. Until re-

cently this information was shared only by a small number of scientists in the United States, Great Britain, and Canada, and the secret could be kept within this narrow circle because no one else had developed the facilities for duplicating the measurements they had made. But of course, as scientific leaders have sought to remind us from the first, the atom knows no national allegiance, and it was therefore only a matter of time until our American "secret" would be discovered by others who would parallel the researches that had afforded us our knowledge—as the French and, more recently, the Russians have apparently now done to a significant degree. When one says that he knows a fact in nature which he intends to preserve as a secret, he means merely that he will not voluntarily reveal his knowledge. Nevertheless the knowledge may be acquired elsewhere. Louis N. Ridenour, himself a distinguished physicist and dean of the Graduate School at the University of Illinois, put the matter this way: "I am saying to you, not that you can not find out what I know, but that you must find it out for yourself, without my help. This may cause you to become annoyed with me, but it cannot keep you in ignorance." [1]

The considerations that bear upon attempted retention of these two types of secrets are different, as is the likelihood of success in the attempt. As to the first type—exemplified by the number of our atomic weapons—Senator Brien McMahon, chairman of the Joint Congressional Committee on Atomic Energy, has strongly suggested that in keeping secret our atomic production figures we "are risking the tested, traditional principles of free and constitutional government," because Congress, being uninformed, "lacks sufficient knowledge upon which to discharge its own Constitutional duties." [2] The number of persons who have information concerning production rates, production quantities, and atomic bomb stock piles is much less than twenty. [3] And Senator McMahon, though he is the head of the Congressional committee which has the responsibility of keeping intimately in touch with atomic energy problems, is not one of them. The issue of whether or not this type of secret should be revealed impressed the Senator as "tremendously important both from the viewpoint of democratic government and from the viewpoint of national defense." A few days after the issue had been raised, President Truman remarked that he deemed it an inappropriate subject for public discussion, an attitude seemingly shared at the moment by most of Senator McMahon's colleagues in Congress.[4]

But whatever may be the merits of matters of that sort (in which scientists' interest is no different from that of all other citizens), the arguments which bear upon them are not the same as those relating to freer dissemination of information having professional significance.

Existing "scientific secrets" are unlikely to remain so for long if anyone is sufficiently interested in duplicating them. Even in the closely guarded realm of nucleonics scientists in England, Denmark, and Sweden have published material that is still classified in this country, while French scientists under Professor Joliot-Curie and his associates Goldschmidt and Kowarski in 1948 successfully produced a chain reaction in the atomic fission of uranium's light isotope, U-235. The French experimental reactor is of much less power than its American counterparts, to be sure, but according to Dr. Joliot it favorably compares with the first American pile (1942) or the first English pile (1947). The French have proclaimed their intention of publishing their research findings without restriction. If this occurs, it is scarcely to be expected that American observations concerning the phenomena of slow-neutron fission will remain unrepeated and unknown. The "atomic explosion" which occurred in the Soviet Union in September 1949 adequately evidences that Russian scientists have achieved a grasp of the subject without awaiting systematic instruction by either their American colleagues or the French.

Americans must constantly remind themselves that the scientific brains of the universe are not providentially concentrated in this country. Recent efforts of propagandists in the Soviet Union to demonstrate that virtually all scientific discoveries were made by Russian nationals have caused merriment in countries where it is not unpatriotic to laugh out loud. American scientists are happily free from this sort of self-adulation. Nevertheless there is perhaps a tendency in uninformed and unofficial American circles almost to match the officially inspired fervor of the Russians. Fortunately for the rest of the world, however, the vaunted scientific superiority of the United States does not derive from some peculiarly national development of human mentality. Many of the ideas, much of the basic research, which have been the solid foundations of American developments have come from abroad. Since the inception of the Nobel awards for distinguished scientific work, thirty-six prizes in chemistry have been granted to Europeans and only five to Americans; of the forty awards in physics, only eight have gone to Americans; thirty-seven prizes in physiology and medicine have been given, of which only six were awarded to Americans.[5] "At present," writes one of our able physicists who himself emigrated from Holland, "the roster of some of our specialized scientific societies reads like the line-up of a Notre Dame football team. In the future, we may not be able to import an Enrico Fermi, whose work was the key to our atom bomb, or a great aerodynamical theorist like Von Kármán, or the outstanding expert on vibrations, Stephen Timoshenko, and many others."[6]

Even in the realms where American technological magic has been regarded as decisive, our debts to other lands are tremendous. It has been said by one distinguished historian, for example, that the resonant cavity magnetron, the revolutionary discovery of British physicists headed by Professor N. L. Oliphant of Birmingham, was "the most valuable cargo ever brought to our shores. It sparked the whole development of microwave radar and constituted the most important item in reverse Lend-Lease." [7] Similarly, the development of the atomic bomb, which so many of us like to regard as a purely American product, would have been unlikely without reliance on the work and ideas of Strassman and Hahn in Germany, Bohr and Frisch in Denmark, De Broglie in France, and many others, including, of course, Albert Einstein. It bears repeating that the men who stimulated this country's interest in attempting to use the Hahn-Strassman discovery of the fissionability of uranium were Enrico Fermi, who had won the Nobel Prize in physics when he was a professor in his native Italy, and Albert Einstein, Leo Szilard, and Eugene P. Wigner, all of whom were mature scientists before they were American citizens.

According to many observers, German scientific endeavors in the period before World War II were enfeebled not only by the racist and political intrusions of the Nazi regime but also by the complacent conviction that German scientists were pre-eminent. This led to abandoning the give-and-take of science; German scientists neither gave of themselves nor strove diligently to learn from the rest. Yet, as events proved, the Germans were far from omniscient and omnicompetent. [8] No doubt the United States, too, can still advance the limits of its scientific understanding by drawing upon the wisdom of others in matters both large and small. Professor Henry DeW. Smyth of Princeton, now a member of the Atomic Energy Commission, tells an illuminating anecdote involving a brilliant young Brazilian, C. M. G. Lattes, who, still in his twenties, has been appointed to a professorship at the University of São Paulo. Dr. Lattes studied at São Paulo and subsequently at the University of Bristol. Then he went to

Berkeley to visit the Radiation Laboratory of the University of California. By applying work he had previously done in connection with the tracks of mesons produced by cosmic rays, the Brazilian scientist quickly discovered that mesons, the forces which hold the particles of the atomic nucleus together, were being produced artificially by the big cyclotron at Berkeley. Until that time the California physicists had been unaware that the cyclotron had been manufacturing mesons for months, though this has subsequently been described as one of the most important events in physics since the war. It may be added, by way of completing this illustration of the international distribution of scientific talent, that the existence of the meson was first predicted in 1935 by Professor Hideki Yukawa of Kyoto University, and that Dr. Lattes while at Bristol was trained by Professor Powell, an Englishman, and Professor Occhilini, an Italian.

Science throughout its history has been strongly marked by coincidences which emphasize how unlikely it is that ideas can be made to flow in narrowly national channels.[9] Chancellor Arthur H. Compton of Washington University, who was one of the outstanding contributors to work on the atomic bomb, received the Nobel Prize in physics in 1927 because of his explanation of the inelastic scattering of light quanta by free electrons. Simultaneously, Peter Debye, now chairman of the Department of Chemistry at Cornell but then a Dutch citizen and professor at the University of Utrecht, was announcing the same conclusions based on parallel researches. American physicists speak understandingly of "the Compton effect"; their colleagues in the Netherlands mean precisely the same thing when they speak of "the Debye effect." In 1949 Professor Edwin M. McMillan of the University of California announced the development and operation of a synchrotron which liberates X-rays of 300,000,000 electron volts and which, it is hoped, will facilitate further research into the splitting of protons and neutrons into still smaller nuclear particles. The "theory of phase stability" that led to devices of this type for accelerating electrons and atomic nuclei to high energies was advanced by Professor McMillan in 1945, when he invented the synchrotron, and in the same year Dr. Julian S. Schwinger of Harvard invented the microtron, another type of particle accelerator. Independently of the American physicists a Russian scientist, V. Veksler, had proposed the same theory for achieving atom smashing. In the summer of 1945 he published in the *Journal of Physics* of the USSR a description of both a synchrotron and a microtron.[10]

Illustrations of this sort of duplication of creative thinking are as readily found in the biological sciences. The analysis of the contagious and septic character of puerperal fever by Oliver Wendell Holmes in this country and Ignaz Semmelweiss in Austria is a century-old tale that still stirs the imagination. It has its contemporary counterparts. In early 1942 an inter-allies group of scientists, co-operating under the auspices of our federal government, developed an immunization technique which so effectively forestalled typhus fever that not a single American soldier died of it during World War II. Their work was not promptly described in the professional journals, lest enemy troops also benefit. Wholly unaware of the completed researches, a second group working independently in a university laboratory duplicated some of the discoveries and published their findings before the Typhus Commission had released the information already acquired. During the war years two governmentally employed groups, who were separately investigating bacterial warfare possibilities, achieved approximately simultaneously the then unparalleled feat of isolating a bacterial toxin in a completely pure form. Their work was not immediately published because of secrecy restrictions. On May 17, 1946, the accomplishment of one of these groups appeared in print for the first time. On that very same day a paper was published by Western Reserve University scientists, wholly unconnected with the bacterial warfare project and uninformed concerning the work there, reporting a similar success with the isolation of a bacterial toxin.[11]

These episodes sufficiently illustrate the impossibility of permanently "keeping a scientific secret" or of precluding others from independently duplicating the most closely guarded researches. They suggest, too, that no particular laboratory is likely at any given moment to possess a monopoly of the scientific competence that makes possible the breaking of new ground. And this would be true as well if all the personnel of all the laboratories of any one country were to be lumped together in a single organization. No country, the United States or any other, is so far ahead of the world at large in scientific attainment that nothing remains to be learned from beyond its own national boundaries.

Unfortunately, the choice of whether or not we shall learn from others does not lie wholly with us. Even if the United States were to embark upon a policy of fully publishing the fruits of scientific work in this country, there is no assurance that all others would pursue the same course. Indeed, the contrary seems probable. The Soviet Union has been even more doggedly secretive and isolationist than the United States. It has rebuffed numerous proposals for cultural and scientific exchanges between the two countries, has virtually forbidden direct contact between Russian scientists and those of other countries, and has frowned upon reciprocal disclosures of research findings even in such entirely nonpolitical matters as the investigation of cancer.[12]

For present purposes, therefore, it must be assumed that there will be no neat balance between outgo of our information and intake by us of others' findings. That may, however, be largely irrelevant. What is now involved is not a species of international bookkeeping, in which purchases and sales are to be recorded. The question to be considered is simply whether restrictions upon the flow of knowledge within the United States may not so gravely impair this country's efficiency that the cost of secrecy will become prohibitive. The issues deserve to be realistically explored without undue moralizing and without supposing a world differently organized from the one we inhabit, that is to say, a world in which international tension and armaments competition will not end soon or, perhaps, ever.

Scientists themselves have not been of a single mind concerning the direction in which our national interest lies. Even though, on the whole, they have not shared the popular enthusiasm for secretiveness as such, scientists have displayed two quite different attitudes toward enforcement of secrecy as a means of maintaining military pre-eminence.

On the one hand, some have asserted that only through unrestricted access to knowledge, in an atmosphere of freedom of analysis and consultation, can science continue to progress. From this standpoint the views of scientists may be summarized as follows:

1. Scientific progress is a prime requisite of the nation's economic and military security. Without it this country cannot keep pace with potential competitors;
2. Scientific progress is unlikely if there is not a full and free interchange of ideas and discoveries;
3. Therefore, national security requires full freedom for scientists and for science.

On the other hand, there are those who believe that since science is not likely to progress except in a democratic environment, which would perish if the Soviet bloc of nations were to dominate the world, the traditional freedom of scientific interchange must be at least temporarily surrendered. Here the position may be summarized in this way:

1. Modern warfare is total war, involving all national resources, both human and material; every activity of every person; every phase of industry and agriculture; and every form and variety of social and political organization;
2. Scientific knowledge bearing upon any of these national resources bears upon the nation's war potential;
3. Therefore, all knowledge must be considered secret and kept under strict security regulations.[13]

Each of these syllogisms presents difficulty. The frightening products of scientific progress immediately reduce one's enthusiasm for entrusting to possibly irresponsible hands a body of knowledge that might be abused. Acceptance of the second approach, on the other hand, would not only prevent transmission of information to potential enemies but would also immobilize our own scientific resources to such an extent that further development might be stifled while more alert countries overtook and surpassed us.

Because the first of the two propositions has run counter to popular belief and emotion while the second has not been palatable even to the most "security minded," there has been continuing search for mechanisms and policies that protect against dissemination of information without at the same time preventing the acquisition of yet more information of a scientific character.

It is noteworthy that American scientists, by purely voluntary self-restraint, have limited the interchange of ideas and information in some circumstances. In the early stages of the work which led to the atom bomb it was the scientists, not the military, who insisted that there be no discussion of efforts by nuclear physicists and chemists to translate theories into performance.[14] Similarly a detailed technical analysis of the subject of germ warfare, prepared unofficially by scientists at the College of Physicians and Surgeons of Columbia University, was suppressed by them throughout the war years and was not published until 1947.[15]

In point of fact, however, self-restraint can operate in only a limited way today, because it has been supplanted by statutory and regulatory commands that rather thoroughly occupy the field. Trammels upon communication between scientists are not measured by individual discretion. Rather they are imposed by official "classification" of data into various degrees of secrecy, which prevent disclosure to unauthorized persons.

This basic type of restriction long antedated the utilization of nuclear fission for military purposes. But since it was the Hiroshima and Nagasaki bombings that underscored the role of secrecy in science, description of the classification process may well be commenced by reference to the Atomic Energy Act of 1946. We turn now to a consideration of the methods whereby a bit of scientific information acquires its status as a "secret." Later, after examination of the mechanics of secret keeping, there will be further discussion of the effects of the process.

Identifying an Atomic Energy Secret

The law that created the Atomic Energy Commission vested it with tremendous authority to bottle up and conceal scientific information. At the same time the statute perplexedly recognized that complete and permanent secrecy would impair, perhaps fatally, the hope of further advance.

The Atomic Energy Act defines as "restricted data" all information concerning "the manufacture or utilization of atomic weapons, the production of fissionable material, or the use of fissionable material in the production of power." So long as data are "restricted" in this sense, they may not be transmitted in unauthorized ways without the risk of grave legal penalties.[16]

Only the Atomic Energy Commission may free information from this statutory restriction by determining that it "may be published without adversely affecting the common defense and security." But Section 10(a) of the Act, while again cautioning the Commission "to control the dissemination of restricted data in such a manner as to assure the common defense and security," expresses a Congressional judgment that "the dissemination of scientific and technical information relating to atomic energy should be permitted and encouraged so as to provide that free interchange of ideas and criticisms which is essential to scientific progress." Thus the Commission has been given the baffling task of balancing two superficially antithetical desiderata—on the one hand, secrecy to assure national security and, on the other hand, freedom of interchange to assure scientific progress.

Failure of agreement upon international control of atomic energy has placed the Commission under unremitting pressure to resolve all doubts in favor of security considerations. While scientists may grumble because, as many believe, the "declassification" of data is too slow, the Commission faces the constant threat of Congressional denunciation if it but slightly disarranges the iron curtain of secrecy. A minor but revealing example occurred early in the summer of 1949, after the AEC on April 28, 1949, had shipped one millicurie of isotope Iron-59 to the Defense Research Institute of the Norwegian government. The declared purpose was to aid a study of "the rate of diffusion of iron in steel at high temperatures." Charging that the shipment of this isotope to Norway might lead to valuable developmental research into the attributes of steel and might thus have a bearing upon military programs, Senator Hickenlooper of Iowa thunderously asserted that the Atomic Energy Commission had been guilty "of a serious breach of responsibility" that involved "potential impairment of our national security." The resulting controversy concerning the shipment of a quantity of material possessing the radioactive equivalent of one one-thousandth of a gram of radium cannot be dismissed simply as a partisan political exercise. Rather, it must be deemed a symptom of a much larger controversy concerning the wisdom of distributing the knowledge gained through scientific research in this country, or of facilitating the acquisition of information by scientists in general.

The true character of the discussion of the Norwegian incident is made abundantly clear by consideration of the nature of the shipment itself. Isotopes have been called "supercharged atoms," a result of bombarding atoms with neutrons. Long before the atomic bomb was devised, isotopes were produced through the use of cyclotrons. With the exception of Uranium 233, Uranium 235, and plutonium, radioactive isotopes are not now thought to be chain-reacting and, so far as research has thus far disclosed, have no utility in the production of power or in the manufacture of atomic bombs. J. Robert Oppenheimer recently told the Joint Congressional Committee on Atomic Energy that even if the isotopes were shipped directly to Russia, he "knew of no way in which this would help them." [17] Their relationship to the bomb is simply that the development of nuclear reactors at the various atomic energy installations and laboratories has multiplied the number of radioactive isotopes available for research purposes. They are, in the words of the Atomic Energy Act, "byproduct material," that is, "radioactive material (except fissionable material) yielded in or made radioactive by exposure to the radiation incident to the processes of producing or utilizing fissionable material." The Commission is authorized by the act to distribute them without charge for research or developmental activity or for medical therapy if distribution will not be "inimical to the common defense and security." Their primary use in research is as "tracers." Since radioactive particles of matter remain identifiable when mixed with other nonradioactive particles of the same description, the path followed by a radioisotope may be traced after it has been mingled with other substances, and thus new light can be shed on the chemical processes of growth and disease, upon the structure of complex materials, and upon the reactions of both organic and inorganic substances in varied circumstances.

Obviously enough, scientific research of any description may conceivably have implications for the military. If the possible were invariably treated as though it were the actual, one would have to conclude that virtually all learning should be kept within this country's boundaries lest it enhance the war potential of some other power. So extreme a position has not as a generality commended itself to the nation's policy makers, for there is recognition that complete confining of scientific knowledge would grievously retard the progress of the United States as well as the progress of its enemies. Yet, as the discussion of the shipment of nonfissionable isotopes has suggested, there is far from complete accord that our national security will in the long run be advanced by facilitating scientific activities throughout the world.[18]

Mindful that the basic question of judgment has no single answer, the Atomic Energy Commission has been distinctly cautious in relaxing the restriction that rests upon scientific data in this field. In the twelve months between November 1947 and November 1948, 1,936 research reports were produced in the laboratories which the AEC controls. Of these reports, over three-quarters (1,567) were deemed by the Commission to contain information that must be kept in a restricted category, and accordingly the reports have been concealed from all but a few selected persons. Two hundred and ten of the research reports related to health and biology; in this group 176 papers, 84 percent of the total, were "classified" and held to be nonpublishable.[19] This is especially interesting because research in the fields of medicine and health have traditionally been "open." Even during the years of active war, the military authorities agreed that publication of new medical findings should be encouraged; the classification of material of this sort was minimized, being confined in the main to limited subjects which were deemed to have immediate battle-front importance or which bore on strategy.[20] Not so in the realms over which the Atomic Energy Commission presides. For many long months after the end of the war, not a scrap of medical research material was declassified. In 1946 it was said that "the entire non-secret literature covering the immense amount of medical work on the effects of radiation and of radioactive poisons on living organisms is to be found in Section 8.70 of the Smyth report. Quoted in its entirety, it is: 'Extensive and valuable results were obtained.' " [21] Even today research work in the biological sciences is perhaps less likely to be declassified than is research in physics, chemistry, and metallurgy, though an encouraging drift in the other direction seems to be presaged by a recent AEC report to Congress.[22]

The figures given in the preceding paragraph suggest the present dimensions of the problem, but they scarcely tell the whole story of the amount of information that remains entombed in the secret publications of the AEC. Not long ago the AEC's Industrial Advisory Group completed a survey of the project. When they finished their work, they commented upon the many interesting and valuable techniques they had observed, the new chemical treatments to protect against corrosion, the instrumentation and plastics and other developments that had grown out of research on atomic energy but had only an incidental relation to it. "We have the impression," added this group of conservative counselors, "that for reasons which are not at all clear, much of this knowledge is still buried in the files and activities of the Commission." [23] It must be borne in mind that the work which now goes forward is a further development of projects which have roots in the past. The record of the underlying researches remains largely unrevealed. Of all the technical and scientific papers that have grown out of atomic energy work, only about 3,200 in all had, as of December 1, 1949, been cleared for release, and these included documents written in the first instance for such varied purposes as oral presentation at public gatherings, publication in newspapers or periodicals, specifications for manufacturing or supply contracts, and so on.

The AEC's Process of Declassification or "De-secretization"

A word should be said here concerning the process of declassification by the Atomic Energy Commission. Slowness in bringing past work to the attention of current researchers is not wholly a matter of policy, nor is it a matter of obtuse obstructionism. In part it is traceable to the scope and the complexity of the task.

Determination that data contained in a research report need not be restricted is a responsibility in the first instance of an official in the establishment where the information originates. If he believes that a paper may suitably be declassified, he must refer it to a "Responsible Reviewer"—one of a corps of a hundred-odd persons, of whom most are specialists in various scientific fields, though a few are individuals possessing an editorial rather than a scientific background. The Responsible Reviewer may decide in favor of declassification or he may deny the clearance sought. When in doubt, he passes the problem to one of four outstanding scientists who are known as "Senior Responsible Reviewers"—W. C. Johnson, chairman of the Department of Chemistry at the University of Chicago;

W. F. Libby, professor in the same department; J. M. B. Kellogg, a division leader at Los Alamos; and R. L. Thornton, professor of physics at the University of California.

All decisions are made in accordance with an officially adopted "Declassification Guide." This document, originally prepared by the Manhattan Engineer District, the Army-administered predecessor of the AEC, has been thrice revised since 1947 conjointly by the authorities of the United States, Canada, and Great Britain. These countries, which shared in the wartime work on the bomb, have pursued identical policies concerning release of the information acquired during the period of their productive partnership. What those policies are, cannot be discussed with precision. The "Declassification Guide" which embodies them is itself a highly restricted document because it lists some sixty categories of nonpublishable information, and thus might possibly serve to identify the types of data having especial bearing upon the production of fissionable materials and weapons.

In addition to moving, via declassification, toward publication of the previously unpublishable, the Commission has taken another important step toward freeing scientific work from restraint. It has defined certain very limited "unclassified areas" in which investigations may go forward and results may be reported without the need of obtaining prior clearance even though they have a tangential relation to atomic energy.[24]

No matter how well intentioned may be the effort to remove secrecy from things which need not be kept secret, the process is a slow one. Ever since the end of the war plans have been afoot to publish a series of technical studies, the "National Nuclear Energy Series," in which would be embodied the research done while atomic energy was still a military project. Some sixty volumes of classified research will ultimately be reproduced for distribution exclusively to project workers who need access to the restricted data they contain. A second and separate group of sixty volumes, each containing about five hundred pages of unclassified research reports, was planned to be given a much larger circulation by being made available to the scientific community at large. As of January 1949 only a single volume, *The Histopathology of Irradiation from External and Internal Sources,* had been placed on sale. During 1949 the book list grew gradually. An additional volume appeared in June, and half a dozen more titles had trickled off the presses by December. A continuing stream, though a small one, may now be expected. Meanwhile, however, enthusiasm for this publishing project has waned. Some of the researches that produced fresh and exciting results in 1945 have been repeated and have been independently published by men who unwittingly duplicated work laboriously completed during the war, and some of the original work that was scheduled for publication in the "Tech Series" has been submitted to regular periodicals by authors who simply grew tired of waiting. Moreover, many of the research papers that are now deemed eligible for disclosure in the "Tech Series" require a measure of rewriting in order to make them publishable. Busy scientists who have long since passed to other activities are somewhat reluctant to interrupt current work in order to refurbish their old reports.

Whatever be the causes, the delay itself has been unfortunate in its effect. The outstanding industrialists who serve as official advisers to the AEC recently recorded "the distinct impression that a vast amount of nonsecret information about the work of the Commission and its predecessor, the Manhattan Engineer District, has never been published anywhere. This type of material can only be made available if the Commission devotes more effort to the task of sorting out the nonsecret from the secret for publication. Frequently this nonsecret information which has not been published anywhere is essential to a clear understanding of that which has already been published in some form." The same body added the conclusion that as to much other information "still classed as secret, the continuance of secrecy is of doubtful value." [25]

Let it be added, to the Commission's great credit, that it reacted positively to this criticism. It appointed a technological working party to search the files of its Patent Branch for matters of industrial use that were unnecessarily secreted there. It stepped up its release of patents and patent applications, thus making available to industry technological information that had previously been concealed. Finally, recognizing that the Russian atomic explosion showed possession of scientific knowledge still withheld from Americans, the AEC in conjunction with Britain and Canada gave renewed thought to releasing rudimentary data concerning already obsolete low-power reactors, as a stimulant of further industrial interest.[26]

How Scientific Data Become Military Secrets

The classification and declassification of information by other federal departments and agencies, notably the military services, are in an even less satisfactory situation.

Power to restrict dissemination of information has not been specifically conferred on federal agencies, though it has long been exercised. The legal authority, so far as it exists, is derived from a general statute having to do with administrative management; it authorizes the head of each department "to prescribe regulations, not inconsistent with law, for the government of his department, the conduct of its officers and clerks, the distribution and performance of its business, and *the custody, use and preservation of the records, papers and property appertaining to it.*" [27] This broadly stated grant, stemming from statutes which trace back to 1789, is the support of today's elaborate classification of scientific data.

Prior to World War II only the War, Navy, and State Departments maintained classification programs that were designed to promote national military security and diplomatic strength, although in late years comparable steps have been taken to assure the security of papers in various specialized fields over which other departments have jurisdiction.

Discussion of the classification programs in general terms is perforce unrealistic. The military services have published skeletal regulations which reveal some of the guidelines but little of the day-to-day practicalities. The published regulations are supplemented by detailed operating instructions which are themselves classified as "restricted" or "confidential." In the nonmilitary departments and agencies there is even less light concerning policies and practices in this general area. Early in 1947 the President directed in Part VI-2 of Executive Order No. 9835 that "The Security Advisory Board of the State-War-Navy Coordinating Committee shall draft rules applicable to the handling and transmission of confidential documents and other documents and information which should not be publicly disclosed, and upon approval by the President such rules shall constitute the minimum standards for the handling and transmission of such documents and information, and shall be applicable to all departments and agencies of the executive branch." The effort to develop a uniform regulation in accordance with this mandate came to grief when, through a news leak, it became known that the State-War-Navy Coordinating Committee had considered placing under security restrictions any information that might prove to be "administratively embarrassing." The resulting outcry and an intensely critical hearing before a committee of the House of Representatives [28] discouraged further efforts to define for all agencies a uniform classification program. The Coordinating Committee itself, which had been created in the first place for quite different purposes and which was only fleetingly concerned with classification, was dissolved as of June 30, 1949.

For present purposes it is enough to describe in broad terms the systems that have developed in the Army and the Navy, which may be taken as representative and which, moreover, are of particular interest because they affect so sizable a portion of the nation's scientific activity. Unlike the Atomic Energy Act, which at the outset places a blanket restriction on all data relating to atomic energy, the applicable military regulations nowhere fasten an embracive classification on any single type of information. Each document is to be classified individually or left unclassified, subject to future change.[29] The four gradations of classification, in descending scale of severity, are "top secret," "secret," "confidential," and "restricted."

The responsibility for classifying documents is highly decentralized and personal. One of the Army regulations reflects a desire that "the least restrictive classification consistent with the proper safeguarding of the contents may be assigned." [30] It is a fair guess, however, that the natural tendency to "play it safe" is almost certainly magnified when a scientifically unlearned person must make determinations which affect the communicability of scientific data. As Dr. Steelman soberly reported to President Truman, the Army's adjuration to avoid too strict classification simply "runs counter to the hard facts of military life. The classifying officer knows that he will never be courtmartialed for excessive precautions, whereas he might be for some error on the side of laxity." [31] Demonstrations of the soundness of this generalization abound. One example will suffice. In 1942, after the British had sought this country's aid in developing special weapons for use in occupied countries, the Office of Scientific Research and Development requested Columbia University to undertake a "study of the corrosion of copper chloride solution." It was hoped that the study would lead to improvement in the so-called "pencil," a simple weapon of sabotage containing a time-delay fuse, already in wide use by the armies of the United States, Great Britain, and Russia. Indeed, many "pencils" had fallen into German hands, and German copies were already being used against the British. Nevertheless, for the better part of a year such stern security restrictions were in force that neither Columbia nor those who were immediately engaged in the studies could be apprised of the purpose of their work. The official historian has mildly noted that "the effectiveness of the group was hampered" by this excessive secrecy.[32] There is no record that the classifying officer's unwise zeal led to embarrassment for him.

Of course the fact that a document has been classified as "top secret" or "secret" or "confidential" or "restricted" does not mean that it becomes invisible. It means merely that it passes out of the zone of easy communicability into a zone where reference to it becomes legally and no doubt psychologically difficult. In the first place, classified information is not readily available to all who might conceivably find it useful, but only to those whose position or work gives them some special claim to it.[33] In the second place, when private individuals do gain access to classified information, they are strongly reminded of their obligation to safeguard it. Thus, every contractor whose operations involve knowledge of military matters because perforce he is given specifications to guide his performance of the contract, is made aware that "disclosure of information relating to the work contracted for hereunder to any person not entitled to receive it, or failure to safeguard all top secret, secret, or confidential and restricted matter that may come to the Contractor or any person under his control in connection with work under this contract, may subject the Contractor, his agents, employees, and

30

subcontractors to liability under the laws of the United States"
—which are then cited at frightening length.[34]

The Declassification of Military-scientific Secrets

In any classification system some provision must be made
for altering or removing an existing classification in the light
of changing events and policy. The Army theoretically permits
a classification to be cancelled by the authority which affixed
it or by any higher authority; and if what is needed is a revision rather than a cancellation, it may be made by any officer
who would have been authorized to give the document its
initial classification. In some especially important matters
there must also be agreement to declassification or revision by
other divisions, including Intelligence and Operations. The
Navy's regulations state that if a document's custodian believes that its classification is insufficiently restrictive, he must
refer it back to its originator or to the Chief of Naval Operations for proper classification. When the need for the original
classification is thought to have passed, the document may be
placed in a less restrictive category by its originator, his superiors, the chief of a cognizant bureau, or the Chief of Naval
Operations.[35]

As might be expected, the urge to declassify does not match
the zeal to classify. The wartime experience of the Office of
Scientific Research and Development is illuminating in this
respect. Here was an organization administered by scientists
and devoted exclusively to scientific work. In security matters,
however, it took guidance from the services. Their classification regulations were accepted and applied without formal
demur, except that the OSRD did seek to avoid assignments
which were classed as "top secret" and which had correspondingly rigid requirements with respect to handling, transmission, and filing. Most of the OSRD research projects were

initially classified as confidential or secret. Once this characterization was applied, it was likely to remain forever. "One
criticism of the OSRD practice which probably would apply
to security precautions generally," wrote an OSRD administrator after the war, "was the persistence of the classification
after the reason for its establishment had ceased to exist. A
periodic review of all classified items would doubtless have
shown many for which the classification could have been
lowered or even removed . . . In retrospect it seems possible
that the saving in time resulting from handling documents of
lower classification would have justified strenuous efforts to
find the time for reclassification at an earlier date." [36] But
"strenuous efforts" are rarely made in this realm. One despair-
ing researcher has casually offered a suggestion that may warrant serious consideration. He has proposed that the classification of any particular scientific data should automatically drop
one notch every six months in the absence of specific action
to reaffirm an existing classification. Thus at six-month inter-
vals a "secret" report would become in turn "confidential,"
"restricted," and "unrestricted" unless affirmative steps were
taken to preserve the limitations upon its circulation. In this
way inertia would lead to ultimate declassification instead of
to retention of unnecessary limitations.

Toward the end of World War II a special problem of declassification arose with reference to the release of the extensive scientific and industrial data that fell into the hands of
American armed forces as they penetrated into enemy ter-
ritory. Acting under his constitutional authority as Commander in Chief, the President determined that these spoils
of war should promptly be released in this country for the
benefit of the American public, always, however, with primary
regard for the omnipresent demands of security. By executive order the President authorized the Director of War Mobilization and Reconversion to take appropriate steps toward

effectuating publication of information about the enemy's scientific and technological advances. At the same time, however, the Secretary of War and the Secretary of Navy were given absolute and final power to forestall release of data if in the opinion of either one of them "the national military security" would be affected.[37] As is customary, the pressures pushing in the direction of revealing what has hitherto been concealed have proved less steady and on the whole less powerful than the characteristic dead weight of declassification authorities.[38] A somewhat parallel situation arose in the OSRD when it faced the problem of publishing the mass of information that had accumulated during five years of scientific silence. The most important phase of the publication program as it finally took shape was a series known as the "Summary Technical Reports." The coverage of these reports was very broad, a circumstance leading at once to their being placed under tight security restrictions which prevented any public distribution. As a result, only 250 copies of the "Summary Technical Reports" were printed, and most of these have been deposited with the Army and the Navy. A small number have been lodged in the archives for possible future distribution or duplication, though, as an official historian unhappily remarked, "the contents are likely to be obsolete before declassification." [39]

II

The Balance Sheet of Secrecy

LIKE most other policies which operate in a complex society, the policy of enshrouding scientific developments in
a cloak of secrecy is neither all gain nor all loss. In this instance, however, there is so wide an understanding of the gain
that the less obvious but nonetheless real loss may be virtually overlooked during public discussions. It is the purpose
of the present chapter to trace the disadvantages of the United
States' position as it has been developing in recent years.

But first it is fitting to restate the objectives of the secrecy
policy. The resolve to try to "keep secrets" was not the act
of perverse or irrational men. It was the act of men genuinely
and patriotically convinced that secrecy would retard the military development of possible enemies. Even though the nation's competitors might ultimately be able to duplicate American achievements, nevertheless the attendant expense, effort,
and delay were deemed to be positive advantages for the
United States. This view is entirely plausible, and the ex-
igencies of the times make it persuasive to most of us. Especially
as to the newer weapons of mass destruction such as the hydrogen or the atomic bomb, the dissemination of information
concerning American discoveries might create perils which
could not subsequently be controlled. Because readier publication of American scientific findings might very well prove

useful to a hostile power, one instinctively applauds secrecy and restraint. No one wishes to place a club in the hands of a potential attacker.

And yet there is another side to the story. If the policy of secrecy is applied with undiscriminating stringency, it may lead to our own ruin. This is overlooked by many who see secrecy as merely a sort of international sanitation. Those who criticize secrecy are often themselves criticized as insincere or ingenuous. This somewhat discourages honest efforts to re-valuate a vitally important policy which bears directly upon national well-being.

No matter how fleetingly unpopular it may be to do so, however, one cannot too often stress that strength lies only in a dynamic rather than a static utilization of resources. The United States may find itself left behind on the road to leader-ship if it contents itself with vigorously marking time. The problem is not one to be viewed entirely as a short-run con-cern. There is more to be decided than whether a momentary hobbling of scientific traffic would be disastrous to the nation. Of course it would not be. Unfortunately, the present issue does not involve restraints of only a moment's duration. It in-volves restraints which have already extended over a con-siderable period of time and which seem likely to continue far into the future unless the balance sheet is reread. The life of a people is long. The effects of a policy on a people must be gauged in terms of future as well as immediate consequences.

The Predictably Unpredictable Uses of Scientific Knowledge

This branch of the discussion may well be commenced by considering the unpredictable course of scientific develop-ment. Who knows what value any given discovery may ulti-mately have? Faraday, when questioned concerning the worth of electromagnetism, countered with another question, "What good is a new baby?" His question suggests the truism that

35

when circulation of knowledge is discouraged, there is an equal discouragement of speculation and experimentation concerning its applications. The implications of data are frequently more important than the data themselves, but of course the implications cannot be pursued if the data are not widely available.

Vannevar Bush has pointed out that many great advances in medical science "have arisen as by-products from such unexpected places as the dye industry"; occasionally a brilliant medical man has created an entirely novel approach to unsolved problems, but more often the steps forward have come about "because other and neighboring sciences were progressing at a prodigious rate, and applications were bound to occur."[1] So it is with most branches of scientific movement. Information acquired for one purpose has proved to have its largest significance in wholly unanticipated ways. Galvani did not have the electric telegraph and the transatlantic cable in mind when he observed that frogs' legs moved convulsively upon being brought in contact with iron and copper; but that observation was the opening phase of the investigations which led to long-distance communication. The present day is equally likely to see dramatic leaps from one body of discovery to another.

During World War II the nitrogen mustards were seriously considered as chemical warfare agents. Chemists at the University of Iowa successfully synthesized and stabilized some forty different nitrogen mustards. Studies of the toxicity and vesicancy of different compounds were undertaken at the University of Chicago. Biochemical studies went forward at the Rockefeller Institute, Johns Hopkins, and Washington University. Pharmacological and physiological studies were carried out at New York University and Yale. All these experiments were directed toward throwing light upon the possible utilization of nitrogen mustards for chemical warfare purposes.[2]

In the course of these studies observations were reported concerning the action of one of the nitrogen mustards on bone marrow and on the lymph nodes. From these observations, which were incidental to the main project and which were reported in the general scientific literature only after the war, grew an important series of investigations of the inhibiting effects of the mustards on malignant lymphoma, such as Hodgkin's disease, for which no treatment had been available. This was scarcely a foreseeable result of what was, in the beginning, weapons research.

The applications of British Anti-Lewisite Compound similarly illustrate the unpredictability of scientific progress. With the outbreak of war in 1939 the British, fearful that Germany would employ gas bombs in its attack upon populated centers, worked feverishly on defensive measures. In 1940 the Department of Biochemistry at Oxford submitted to the British Ministry of Supply a secret report concerning a compound that would prevent the blistering effect of the World War I arsenical gas, Lewisite. This compound, known for security reasons simply as OX No. 217, came in time to be called BAL (British Anti-Lewisite). In 1945 the discoverers of this important antidotal agent were at last permitted to publish a brief description of their findings, including the chemical structure of BAL and its mechanism of action. Within a year BAL had been put to successful clinical use in treating arsenic poison complicating the therapy of syphilis and in salvaging the lives of persons who had taken mercury with suicidal intent; subsequently it was found useful in overcoming gold poisoning contracted in the course of arthritis therapy.[3]

Chancellor Arthur H. Compton recently recalled that "fifty years ago we knew already that X-rays were useful for 'seeing'

37

through objects, such as the human body, which are opaque to ordinary light. It could not then be predicted that X-rays would become a powerful weapon in the fight against cancer. No one could foretell that studies with X-rays would reveal the electron, and with this discovery give us eventually the radio and a host of electronic devices. Such unforeseen developments are the result of every great discovery."

How likely is it that similarly important "unforeseen developments" will grow from the release of atomic energy if the free flow of knowledge about it is persistently blocked? The answer to that question is suggested by another member of the Compton family, Karl T. Compton, until recently the chairman of the Research and Development Board of the National Military Establishment and previously president of Massachusetts Institute of Technology. Dr. Compton, testifying before the Senate Committee on Military Affairs in 1945, drew some interesting lessons from the use of internal combustion engines in the airplane, the automobile, the tank, and the bulldozer. "Suppose, about the time when most of us were boys, and the automotive engine was relatively in its infancy, some agency like the War Department had conceived the idea that this might be very useful as a future military development and had clamped down the imposition of secrecy in the further study of high-octane fuels, metallurgy, thermodynamics, and engine design, and all other features which have to go to build the most efficient possible engine. These conditions of secrecy might have involved a prohibition against doing work in this field without a license and against any discussion with other workers in the same field except by Federal permission, and no right of publication of results unless this commission thought that they would be of no aid to any foreign government. We can easily see what the results of such a policy would have been. Our own development of the automotive engine and the great automobile and aircraft business would have been greatly retarded in this country. Other countries operating without such restrictions would have forged far ahead of us. . . . In a similar way, with any development of an important new field of science which may have important practical application for either peace or war, it seems to me that our first consideration for national economy and national security must be to handle this development with a minimum of inhibitions and a maximum of assistance and inducements." [4]

The Compartmentalization of Scientific Work

In a very direct manner the concentrated effort to "keep secrets" ignores what has just been said about the unpredictable ways in which scientific data prove their significance. A central feature of much secrecy administration is "compartmentalization" of the work that is done in various areas. Secrets, it is thought, are most likely to remain so if they are known to only a few people. The less a man knows the less he can tell, even if he is actively disposed to violate the confidence that has been reposed in him. To minimize what any one person may be able to tell, the secrecy administrators have evolved the homespun security principle that he ought to be told only as much as may be necessary for him to get on with his immediate job. And so it is that scientific labors come to be done in separate compartments, which tend to limit the interchange of knowledge.

From the first the Atomic Energy Commission has been committed to a compartmentalization philosophy, though, inconsistently, there happens in fact to be considerable freedom of interchange within the Los Alamos laboratory. The Commission recently reported that "no person receives more classified information than that needed for the performance of the particular tasks entrusted to him," a restriction which, as the Commission glumly acknowledged, "may work against progress since often one person or group will be in possession of information of great value to others." [5]

The soundness of this observation is fully attested by experiences in comparable areas of scientific endeavor. It is recorded, for example, that at the outset of work in the microwave radar field efforts were made to maintain limits upon the amount of information given each group. Men who worked on separate facets of a single problem were not apprised of their colleagues' efforts, and, indeed, did not even know at times that they had colleagues. This was especially true with reference to the cavity magnetron. The invention of this transmitting tube basically affected the whole project. Nevertheless, men who were assigned to work on a modulator to energize the tube were in the beginning denied knowledge of its design. But progress was so slow, there was such inefficiency and such duplication of research, that the policy was soon abandoned. By the time the war ended, the Army was the publisher of a radar magazine with a circulation of more than 12,000, for it had become apparent that "secrecy cost us in efficiency more than it gained us by keeping the enemy in ignorance." [6]

The inefficiency of compartmentalization of work—or, more accurately, fragmentation of knowledge—is threefold.

First, fragmentation so narrows the range of expertness that effective utilization of scientifically trained manpower is badly hampered. This country's slowness in World War II in developing fire control with radar for the Navy's long range anti-aircraft guns and main batteries is illustrative. At the beginning of the war our Navy was superior to others in respect of these phases of fire control. The work on fire control was, however, very tightly restricted. When war came, the Bureau of Ordnance was "somewhat unreceptive to new technical groups, which might seek to enter the field"—in part, at least, because "the operation of security regulations had prevented other groups from gaining the intimate knowledge of naval fire-control policies to qualify them as 'experts.' " The achievement of results had to be postponed until this lack could be overcome.[7]

Second, compartmentalization prevents full utilization of work that has already been successfully accomplished. The various national laboratories that are engaged in research bearing upon atomic energy, for example, believe that they frequently repeat work that has been completed at Los Alamos, especially in the field of chemistry. The head of a major division there has vigorously asserted to me, "Too damn much is being declassified"; perhaps as a result of his conviction, information flows to Los Alamos from the other AEC projects without a correspondingly strong return current because he refuses to lower the barriers as readily as do his colleagues elsewhere. Within the Los Alamos laboratory itself there is said to be no compartmentalization; as Dr. J. H. Manley, its Technical Associate Director has said, "In the new and strange field in which this laboratory operates, ideas of value may not necessarily always come from the individual who is supposed to have them, and a free flow of problems and information among the senior scientists is important in maintaining progress." [8] Until recently this recognition of the costs of compartments extended only to the limits of the mesa on which Los Alamos stands. Of late there have been manifestations of readiness to concede that men in other AEC installations have something to learn from Los Alamos and, in turn, to teach it.[9]

Third, compartmentalization necessitates frequent duplication of unfruitful research. The third of these may be even more important than the others, for assuredly one of the highest functions of scientific research is to discover the unpromising approaches and to mark the blind alleys that do not lead to truth. "A research program," it has been said, "is never a failure. Every incident in its history will prove to be an edu-

cational factor in the next investigation undertaken." [10] Sir
Alexander Fleming, the discoverer of penicillin, had somewhat the same thought in mind when he remarked recently
that every research man knows "the weary months spent working in a wrong direction, the disappointments and the failures.
But the failures may be useful, for when properly studied they
can lead to success." [11] The trouble with fragmentation of
knowledge is that it shuts off awareness of the failures and
thus forecloses proper study of them by those who might profit
from them.

All along the line, in truth, compartmentalization prevents
one scientist's learning from another in the traditional way.
The AEC seeks to minimize this difficulty to some extent by
circulating among its various installations and contractors a
title-and-author list of all classified reports, as well as a publication called *Abstracts of Classified Documents,* in which the
contents of new reports are briefly identified. But this is far
from distributing the classified documents themselves, nor
does it assure that work in progress will be facilitated by opportunity for direct observation and personal contact between
persons whose primary assignments may differ, though they
may have much in common in respect of some subsidiary aspect of their researches. This point is well illustrated by a
paragraph in the findings of the AEC's Industrial Advisory
Group, to which earlier references were made. The Group in
its report to the Commission spoke of the need of increasing
the contacts between industry and the Commission, and in
this connection mentioned a member of the Industrial Advisory Group who is himself "in charge of an important specific
industrial research and development project. Among the
knotty unanswered questions in his project is one relating to
the type of coolant to be used. During our survey, he observed,
firsthand, a unique process that was being worked on in one

of the Commission's laboratories to solve a problem which was also related to coolants. The Commission's work immediately suggested to him a new avenue of approach to his own special research problem. He remarked at the time that even had he read about the Commission's investigation in the technical journals, the chances are that he would have missed the connection with his own investigations. Direct personal contact with the work in the atomic energy laboratory gave him the concrete experience necessary to see a relationship that he would otherwise have missed." [12]

Such incidents as this make possible "the massive forward movement of technology." If the erection of barriers between compartments prevents this type of experience, the forward movement will assuredly be at a slower pace. For as this episode suggests, the boundary lines of compartments are unreal and unfunctional. Because of the ramifying significance of particular ideas or technical improvements, the happenings in one compartment may have vital interest far beyond its confines. Few major problems of modern science can be neatly labeled and assigned for solution to a single specialist. As the Director of the Atomic Energy Commission's Research Institute at Iowa State College stated the matter in addressing the Electrochemical Society, "It is possible to design reactors in many ways, and the problem of design in each of these reactors requires the combined efforts and knowledge of almost all kinds of scientists and engineers. Basic discoveries in all the fields of physical chemistry, metallurgy and engineering will have to be drawn upon to make the practical applications, and almost any scientific fact in these fields may prove useful in the practical applications of atomic energy." [13]

Finally, compartmentalization and fragmentation take no account of the needs of those who carry on their work outside the area of secrecy. Matters that have been touched upon

43

within guarded laboratories and in classified documents often have direct importance for activities but slightly related to secret enterprises.

An interesting specific example of this was observed recently at Brookhaven National Laboratory, one of the major research facilities connected with the Atomic Energy Commission. During 1948 and 1949 there was under construction at Brookhaven a new nuclear reactor, the elements of which are within the zone of highest secrecy. At the same time there was being built at Brookhaven a new particle accelerator, a great proton-synchrotron dubbed the "cosmotron," capable of accelerating protons to the velocity of perhaps three billion electron volts. Data related to the cosmotron were not "classified," because the principles which are expressed in the cyclotron, the synchrotron, and like devices are already well understood abroad as well as at home. Those who were responsible for designing the Brookhaven accelerator were dissatisfied with the protective shielding which, used in conjunction with earlier machines of this sort, had guarded the operators against the danger of overdoses of radiation. They felt that a more complete safety device should be installed. The protective shielding around the reactor, or atomic furnace as it has sometimes been called, is said to be highly perfected. But its specifications could not be disclosed without minutely compromising the secrecy that envelops the production of atomic energy through nuclear fission. As a consequence, those who had the Brookhaven accelerator in charge independently developed shielding techniques which they felt were adequate to their needs.

The costs of this sort of duplication can perhaps be measured in terms of time and money, but never in terms of what might have been accomplished if brains had been free to work on the problems of the as yet unknown, instead of on problems which had previously been solved by others. This was perhaps the thought of the Hoover Commission's "task force" which dealt with national security when it reported in 1948 that the Federal Government "is not getting full value" from its billion-dollars-a-year investment in scientific research and development.[14]

Interestingly enough, the inability to profit from another's thinking cuts both ways. A scientist who is engaged in a secret undertaking may be limited in drawing help from others, even though the data or ideas he wants are wholly nonsecret, because the nature of his questions might possibly suggest the direction of his researches and might thus lift a corner of the veil of secrecy. A senior physicist at Los Alamos, for example, recently acknowledged that he is frequently slowed up in attacking a problem by his inability to consult the recognized leaders in that field. Where formerly he would merely have written to one of his professional peers or spoken to him in an informal way, he is forced by secrecy considerations to delve through all the man's published works, and even then he may fail to find what he needs.

Among the causes of the decline of German science in the nineteen-thirties was a growing tendency to carry on researches in an atmosphere of secrecy. Americans who traveled abroad in those days were shocked to find that German laboratory doors were locked—not, be it added, because of governmental edict, but because colleague distrusted colleague and feared that credit for ideas would be stolen. "In Germany," it has been asserted, "scientists never sat around tables together swapping their experiences of trials and errors, telling of how their work was going, asking each other for suggestions." [15] It was precisely this uncommunicativeness which helped retard research and which made for inefficient employment of trained manpower. Yet, as has been seen, American insistence upon fragmentation of knowledge will perforce have the same ultimate effect upon progress here since it will inhibit the ex-

45

changes of scientific ideas and the stimulations that come from a comparison of experience.

This is not a purely speculative statement. It has been frequently remarked, for example, that at Oak Ridge, when every moment counted, related groups worked diligently on the same problem without the slightest awareness that there was duplication of effort. It is said, too, that because there was, and still is, a tendency to be especially secretive about information acquired at Los Alamos, the scientists at the gaseous diffusion plant at Oak Ridge (K-25) were at one time unintentionally exposed to great hazard. The staff at K-25 was uninformed concerning the critical mass of the uranium isotope, that is, the amount which will produce an explosion or a deadly burst of radiation. A possibly apocryphal but widely repeated story tells of a visitor from Los Alamos who discovered quite by accident that at one place in the plant the accumulation was approaching perilously near the critical point. By violating security regulations, he was able to give the Oak Ridge staff the information that averted disaster. Few examples so dramatically reveal the disadvantages of compartmentalization; but in terms of retardation of further research, the reported instance is of lesser significance than the daily accumulation of unspectacular delays which remediable ignorance causes.

It is especially disturbing to reflect that the practice of compartmentalization is continuing in this country despite the freshness of observation concerning its demerits during the past war. The National Defense Research Committee and the Office of Scientific Research and Development from the very beginning accepted the policy, initiated by the military, of compartmentalizing information on the grounds of security. This led to incredible difficulties in carrying forward the research upon which the success of our arms depended. One important research project, for instance, involved inquiry into the effects of various types of projectiles upon structures. The members of this research group, who were students of the defensive properties of concrete and steel, were purposely kept in ignorance of the outcome of tests of the performance of shaped charges against concrete, and were long blocked in efforts to learn the results of projectile firings against reproductions of German pillboxes. Difficulties like this led an official recorder to conclude that "more harm in arresting research and development was done by this compartmentalization of information than could ever have been done by the additional scrap of information that the enemy might have picked up by a more general dissemination of knowledge." [16]

President Irvin Stewart of West Virginia University, executive secretary of the National Defense Research Committee before the war and subsequently the deputy director of the Office of Scientific Research and Development, has acknowledged that compartmentalization made for inefficiency: "In theory," he writes, "the Committee members and later the office of the Chairman had the responsibility for seeing that information crossed divisional lines whenever research would be speeded thereby . . . Unfortunately, however, there were cases in which information in the possession of one division of NDRC was not known to another division, although it would have been very useful to the second division." [17] If barriers had not occasionally been informally and selectively ignored by some of the working scientists, there is reason to believe that many wartime advances would have been delayed if not eliminated. Especially in view of the fact that there appear to have been no seriously indiscreet disclosures of information by American scientific personnel throughout the long years of the war, Dr. Stewart believes "in retrospect that compartmentalization of information to the extent practiced was not in fact needed," though he notes as a high probability that compartmentalization made the military men "more

47

willing to entrust their classified information to the NDRC during the early period when the ability of the organization to keep secrets had not yet been demonstrated." [18]

The serious misdeeds of a single naturalized Briton, Klaus Fuchs, and the subsidiary faithlessness of several petty scientific workers in this country should not be allowed to obscure the realities. Fuchs's perfidy, exposed in 1950 by diligent counterespionage, was not a reflection of an occupational characteristic. On the contrary, every available record emphasizes that his behavior was aberrational, unrepresentative of and uncondoned by the scientific community of which he was a part. By this time the scientists' acceptance and performance of responsibility should successfully have overcome the early doubts of the most skeptical military officers, though there seems to be a calculated effort in some Congressional quarters to arouse concern about the "reliability" of scientists as a group. A similarly suspicious attitude on the part of the Japanese army and navy led to rigid and continuing compartmentalization of scientific endeavors in that country, and this, according to an authoritative historian, significantly contributed to the relative lack of scientific progress in Japan during the war.[19]

Here it is pertinent to quote the words of Joseph C. Boyce, now of the Argonne National Laboratory and the official recorder of this country's work in fire-control equipment, proximity fuzes, and guided missiles: ". . . all too often the development of the various components of a guided missile was given to independent groups in the vain hope that the components so developed would function properly together. Unfortunately this tendency still persists in some quarters. Security is usually quoted as the justification for this procedure. Experience of this war has shown considerable parallelism in the independent development of new weapons in various countries. This is to be expected since the fundamental

48

scientific and engineering principles are available to all nations. Security is wasted if a new development comes too late. Fortunately for us, the Germans and the Japanese made this sort of mistake more frequently than it was made in this country. But enough instances occurred here to waste valuable months." [20]

It is comforting, in a way, to know that someone else made the same mistakes we made. The comfort vanishes if we discover that those very mistakes are to be continued as a matter of policy not only by the services but also by the Atomic Energy Commission.

Loss of Criticism

Compartmentalization and secrecy not only prevent exchanging the information and the hunches that expedite research. They also prevent objective appraisal of the work in progress. Scientists who are constrained not to talk about what they are doing fail to receive the vigorously honest criticism which may save many a false step or which may lift an experimenter's imagination beyond its present limits.

Here we must distinguish between secrecy that is imposed for the very purpose of stifling criticism and, on the other hand, the stifling effect of secrecy which is imposed with wholly different objectives in view. The use of secrecy restrictions to avoid embarrassing disclosures is certainly not unknown. During the last war, as many witnesses have affirmed, mistakes were often concealed by classifying as secret all information which bore on them, and at times, indeed, controversial subjects which had military implications seemed almost automatically nondiscussable because of "security considerations." Even the Atomic Energy Commission, which has often professed a desire to furnish the fullest possible measure of information to the public at large, has not eagerly published what would embarrass it. For example, it was not until October 1948 that the Commission made an apparently routine announcement that former Supreme Court Justice Owen J. Roberts and all the other members of the AEC Personnel Security Review Board had resigned. In fact, the members of that board had resigned in a body during the summer, in large part because of dissatisfaction with the Commission's actions on its recommendations. When the announcement was finally made, of course the surrounding circumstances were not recounted, nor was it revealed that announcement would have been withheld indefinitely but for the fact that a committee of scientists had arranged to confer with the Commission about security procedures. As one Commission official has said, "While it did not embarrass us to hold back news of the Roberts board's resignation, it would have been awkward to talk about the board as though it still existed. So, the day before the conference, we set the record straight." [21]

This sort of misuse of "security" occurs in scientific matters too. Early in the last war, for instance, a scientific unit studying structural defense and offense tested some concrete structures by dropping various general-purpose bombs of the then design. The tests revealed drastic defects in the bombs rather than in the structures. But when it was proposed that the observations and photographs that substantiated these defects should be made available to the British, who also had a considerable scientific interest in the subject matter, delays and difficulties suddenly arose. Eventually the scientists' information was communicated, but not until effort had been expended in persuading the military that great harm might result from unwillingness to learn from failure. [22]

In sum, secrecy may be a device to conceal ignorance and error as well as knowledge and success.

But in the present context it is not proposed to discuss intentional flouting of the principle that the opportunity to scrutinize and criticize is the public's chief protection against governmental incompetence, dishonesty, or abuse. We are discussing, rather, an unintended by-product of scientific uncommunicativeness, namely, the inability to assess and perchance to assist work the content of which is kept secret. A prominent Cornell physicist who serves from time to time as consultant to a government-supported laboratory in which much secret work is done summed up the matter recently by saying, "Since nobody knows what these people are doing, they are not kept on their mettle. They tend to stagnate for want of honest competition. Secrecy is creating a new class of scientists, inbred and aloof." Who can say whether the projects that are chosen for extended research are chosen wisely? Who can say that they are carried forward in the most effective manner, or that the conclusions derived from them are beyond question? When research is open and its results are published, scientists throughout the country, throughout the world, promptly repeat the experiments in their own laboratories, checking and confirming the published results and computations. Verification of this sort is of course impossible when the results of research are concealed. There is no reason to suppose, however, that secret research is flawless. On the contrary, Dean John R. Dunning of Columbia, a well-known contributor to our wartime scientific endeavors, has asserted that much of the research work done during World War II has subsequently been shown to have been faulty in method or findings.

Mindful that the objective judgments of dispassionate outsiders may be helpful to those who are deeply engrossed in research, the government frequently engages advisory committees or individual consultants to examine particular problems. Thus, for example, a board of eminent medical scientists has toured the research centers of the armed forces, with a view to evaluating the projects which they have launched. But the trouble with this sort of thing is, simply, that it is not continuous. Professor Smyth, author of the famous report, *Atomic Energy for Military Purposes,* remarked before he became a member of the Atomic Energy Commission in 1949 that it is impossible for an outsider who is only occasionally abreast of what goes on in an AEC installation to know whether it concentrates on fruitful lines of inquiry. Just recently confirmation came from the members of the Industrial Advisory Group, which had been established under the Atomic Energy Act to help develop a program for full utilization of the nation's industrial and research capacity. After more than a year's work, during which it was given access freely to all necessary documents, personnel, and installations, the Industrial Advisory Group emphasized in its final report that "despite the excellent cooperation afforded by the Commission, one of the serious obstacles in making our survey arose out of burdensome security regulations. Difficulties in connection with clearances, the complicated mechanics of arranging for access to people and installations, the elaborate procedures for the safeguarding of notes and documents, as well as other secrecy restrictions, together constitute a formidable impediment to any attempt to study and understand the enterprise." [23]

The Psychological Consequences of Secrecy

The matters which have thus far been discussed have dealt mainly with objective, impersonal consequences of secrecy in science. The subjective aspects of the matter also deserve comment. One of the least tangible and yet perhaps most far reaching of the costs of continued secrecy is its psychological impact on those who deal with classified data.

It is of course perilous to generalize concerning human motivations and human reactions. To say that a number of men are scientists is not to say that they have lost their diversity. There is no single type of scientist and, as a corollary, there is no single response to secrecy. Yet it is possible to advance some plausible hypotheses concerning the state of mind of many of the persons upon whose insight and skill we depend for continued scientific advance.

It is known, to begin with, that financial considerations rarely induce embarkation upon a scientific career. Scientists as a group in our society have not been highly paid. They have found their satisfactions elsewhere. In 1947 the National Opinion Research Center of the University of Denver studied the attitudes of an objectively selected cross section of American scientists. Those who were interviewed were asked, among other things, to describe the special attractions they found in their careers as scientists. Intellectual and temperamental satisfactions, along with the social value of the work done, dominated all other things mentioned. Only one percent felt that the economic rewards or the security of a scientist's career made it attractive. On the contrary, nearly four-fifths of the whole group thought the scientist's rewards in money and prestige were so slight that no man should enter upon a scientific career in order to reap them.[24] Scientists remain at their tasks because, in the main, they are excited by the search for a particular kind of truth. This sort of excitement has been sustained by a professional fellowship, scattered yet tightly knit. Men who have engaged in research testify with near unanimity that exchanging ideas and data with others has been invaluably stimulating, not only because it advanced the work in hand but perhaps even more because earning the respect of professional peers has been an incentive to achievement.

Today the exchange of ideas is discouraged by constant stress on maintaining security of information. Men whose work involves access to restricted materials tend to avoid discussion of their activities except with their immediate associates. Scientists who work in the isolation of remote installations like Los Alamos have recently been encouraged to attend scientific and engineering meetings lest their laboratory researches

53

suffer. But once they are there, they are not encouraged to converse with fellow scientists who do not "know the secrets." They are reminded that inferences can possibly be drawn from what they have left unsaid as well as from what they say. They are warned by the Atomic Energy Commission's Office of Security and Intelligence that even when they are dealing with wholly unsecret matters, nevertheless what they say or write may be "flavored" by their memory of classified data.[25]

When a scientist must be mindful not only of his facts but of his flavor as well, it is understandable that reticence governs his intercourse with the rest of the scientific world. It is difficult to know what can be said and to whom it can be said, for even a scientist who has been "cleared" for access to secret data is not by virtue of that fact alone entitled to unrestricted access; as has been seen, he is entitled to have access only to the data he knows he needs in his own work. Avoidance of discussion becomes the comfortable and perhaps even the necessary course in these circumstances. Thorfin R. Hogness, one of America's great scientists who heads Chicago's Institute of Radiobiology and Biophysics, told the convention of the American Veterans Committee on November 25, 1949, "Most men who were once associated with the atomic bomb project and are now cleared as consultants never ask questions from those now engaged in this work. If they did so, they might be regarded as snoopers. Such is the atmosphere created by secrecy."

An outstanding university professor who serves the Los Alamos laboratory as an adviser each summer recently illustrated the reverse side of the coin by remarking, "When I leave Los Alamos, I turn off like a faucet that part of my mind which dealt with my work there. I do not think about those problems at all until I go back the next summer. This is inefficient, of course, but it is the only way I can be sure that

54

classified material will not find its way into my discussions elsewhere."

So it is that the demanding pressures of secrecy make themselves felt in the behavior and the temperament of those who work in the twilight. Gossip, it used to be said, was the lifeblood of science. Today it is taciturnity rather than gossipiness which is enforced upon scientists as a group trait.

Effects of Secrecy on Recruitment and Training

No matter how large may be the appropriations for research and development in "restricted" fields, they by themselves can produce no work of value. The level of achievement will be determined by the quality of the men and women who can be persuaded to use the appropriations. Experience at hand shows that many well-equipped scientists are reluctant to be subjected to the devices already discussed whereby knowledge is fragmentized and its circulation forestalled. Those who agree to work under the restraints do, of course, unhesitatingly observe them. It is likely to be increasingly difficult, however, to recruit additional strong scientists into laboratories that the government dominates through secrecy controls.

The pinch of this problem has already been felt by the armed forces. "It is disturbing," says a recent report to the General Staff, "that so few professional scientists find a permanent military career attractive at a time when the research and development budget of the Services is at an all-time high for a period of peace." [26]

A similar problem affects civilian agencies. Consider the case of the Los Alamos laboratory. It has often been characterized as the best-equipped installation in the world for research in physics, nuclear chemistry, and some areas of biology. It houses, along with all the more conventional equipment, two nuclear reactors devoted to research rather than to large-scale production of fissionable materials, a cyclotron, a betatron, a Cockroft-Walton accelerator, a Van de Graaf accelerator of 2.5-million-volt power, and another Van de Graaf of 12-million-volt power under construction. The salary scale for those who work in this magnificently supported laboratory is higher than that of most universities. A staff member has no teaching burdens, but can devote all his time to research, without fear that lack of funds will block the testing of his ideas. Nevertheless the Atomic Energy Commission has sadly acknowledged that it has not yet persuaded an adequate number of qualified persons to enter the scientific paradise its funds have built.

One must avoid an oversimplified explanation of this sort of difficulty, which is by no means limited to the atomic energy program but runs throughout the research activities of the government.[27]

One cause of reluctance to enter government laboratories may very possibly be the "fear of smear"—the fear that one's reputation or at least one's peace of spirit may be impaired by irresponsible persons, in and out of Congress, who make their major appeal to minds befogged by misconceptions concerning "secrets."

The impact of this factor upon recruitment is, of course, difficult to measure. Vannevar Bush, former chairman of the Research and Development Board of the National Military Establishment, has expressed to me his belief that there has been no impact at all. But there is a respectable body of opinion to the contrary, including that of Dr. Bush's successor, Karl T. Compton, who, discussing "this great furor about possible leaks of secrets," has said: "All of us concerned with progress in military research know that the results of this publicity, and some procedures of official investigating groups, have seriously impeded our progress toward security through scientific advancement"; even the taking of consciously calculated risks that confidential data might pass into unauthorized hands "might be better than creating by law or public opinion conditions which make employment so unattractive that top-flight scientists and engineers go in more comfortable and usually more rewarding directions." [28]

While one can scarcely be dogmatic about the subjective reactions of potential recruits who have simply declined to be recruited, one may assume with fairness that the conduct of the House Committee on Un-American Activities, especially while under the chairmanship of former Representative J. Parnell Thomas, has not actively encouraged persons to seek a career in government-sponsored research. In all likelihood, however, neither demagogy nor ignorance [29] would, alone, make it impossible to attract able scientists. The fear of embarrassment is merely an added discouragement to recruitment rather than a basic explanation of its failure. An official report to the President in 1947 suggested that the two factors chiefly responsible for making the Government's research program somewhat unattractive to scientists are "(1) the heavy concentration on military subjects, and (2) the minor emphasis commonly given to basic research." [30]

The psychological basis of the first of these is easily perceived. Even though the development of improved military mechanisms may be of great importance in a world from which war has not been excluded, the objectives of military research are negative and destructive. Many men who have been trained to think of science as a means of creating good by revealing truth no doubt find it distasteful to readjust values and redirect emotions, as may be required of one who devotes his energies to preparing for war in time of peace.

More important than this, however, according to the Steelman report, is the fact that "the secrecy and censorship which accompany much military research and restrict publication of results make for a competitive handicap in recruiting and retaining the best scientific minds for the Government's military program." [31] Echoing this opinion the Atomic Energy Commission recently observed that "staffing of atomic energy projects is hampered so long as there is feeling on the part of many scientists that employment in the atomic energy program precludes their working on any but 'classified' research projects with consequent denial of general publication." [32]

Professional tradition has long bound the scientist to publish his work for the benefit of and for testing by the rest of the scientific world. "The cumulative nature of scientific knowledge," writes Nobel Laureate Rabi, "puts the scientist in such great debt to the past and to his contemporary colleagues that his responsibility to present his results can hardly be honorably evaded." [33]

For the younger man, this tradition is reinforced by self-interest. It is through publication that an as yet unrecognized man establishes his claims to eminence. Universities and other employers of scientists almost invariably request a job applicant to furnish a list of his writings. A scientist who has been allowed to publish nothing may be able to present glowing reports by his former supervisors; but these are rarely as persuasive as the printed records of his own past labors. If the labors have been secret and if disclosure of their results is prohibited or discouraged, the normal path to professional preferment is blocked, and this is a possibility which an ambitious man must take into account in choosing the work he will do.[34]

Avoidance of "classified" researches has been a felt reality at the Brookhaven National Laboratory. This outstanding laboratory, located on Long Island, is sustained by AEC money, but is administered by nine eastern universities, incorporated for this purpose as Associated Universities. Recognizing that the teaching staffs and the students of colleges and universities are capable of making great contributions, Brookhaven has encouraged their participation in its fundamental nuclear and radiological research. The laboratory is an ideal field establishment for training graduate students and is a pleasant place in which faculty members may pursue their investigations while enjoying the company of colleagues from all over the country. The work to be done at Brookhaven is mainly of an "unclassified" character, but some of it is "restricted" because it involves the uranium-graphite reactor as a research tool. According to Dr. Leland Haworth, the director of Brookhaven, the qualified men at that installation have so great an antipathy to secrecy that research of large importance in the classified area is being neglected in favor of less interesting subjects that can be discussed without restrictions.

The distaste for entering the darker portions of the scientific hinterlands has been manifested in yet another way. Despite the "glamour" of working in such new and highly publicized fields as radiochemistry and radiobiology, many of the most promising students choose other specialties less hedged about by secrecy. This observation has been made by professors in widely scattered institutions. Although an absolute generalization would be unwarranted by the evidence at hand, there is a fully justified fear that many possessors of brilliant minds will exclude themselves from future research in these important realms. One point which the ablest students have stressed is that the radiobiological or radiochemical work they may undertake to do outside the classified laboratories may prove to be merely a duplication of research that has already been done inside them. They prefer to labor in the light, where they can distinguish between tilled and unplowed ground.

So far as training the scientists of the future is concerned, however, the retention of secrecy poses graver problems than the occasional reluctance of an able man to receive training in radiobiology. The real danger of secrecy in this respect is that to some extent it prevents advanced training altogether.

This danger has two aspects. At the outset we must note

that some sorts of schooling, especially in the nuclear studies, entail the use of machinery that few educational institutions can afford to operate safely. The manipulations required for processing radioactive materials, for example, can be learned only in elaborate installations that are rarely found in universities. The limitations of academic resources therefore require that some of the training in these fields be done in laboratories which the Government controls. The introduction of research students into these laboratories involves a complex employment system devised to safeguard "security." All elements of this system, it has been observed, have discouraged able candidates from entering the research training program.[35]

The second aspect of the danger that effective training will be prevented was discussed by Henry DeW. Smyth in the autumn of 1948 in an address before the University Club of New York. Then, as now, one of the world's best-informed men concerning uranium fission, Dr. Smyth was chairman of the Physics Department at Princeton University. But much of his information had to be withheld from his students. He was not allowed, for example, to tell them how many neutrons are given off in uranium fission. How then, he asked, could the current crop of students learn the fundamental facts on which new engineering plants for the use of atomic energy must rest? How can the scientists of the future be given the insights they need to work on problems of atomic development which baffle the scientists of the present?

These questions raise an issue related to but different from the suggestion previously made, that restrictions upon communicating scientific and technological data threaten to freeze rather than free the limits of knowledge. The issue now raised is whether the formal education of a new generation of scientists will have to be confined to subjects in which secrecy regulations do not inhibit discussion between teacher and pupil. The great Fermi was speaking not long ago of his course in nuclear physics at the University of Chicago: "I would have liked," he said, "to give my students a certain background to the work in atomic energy. I have a fair notion of what is classified and what is not classified, but still the feeling that I would have had to weigh my words very carefully—I could have been asked embarrassing questions, and I would have been faced with the choice of either telling a student in the open classroom, 'I am sorry, my boy, but this is something that I am not allowed to answer.' And just this uneasiness drove me to stay off the subject. Now, I do not think my lectures would have been extremely effective, but there you have some 50 boys or so who have lost that chance to acquire training in atomic energy problems." [36]

Philip McC. Morse, former M.I.T. physics professor who served for a time as director of Brookhaven National Laboratory and now directs the Defense Department's weapons evaluation group, says flatly, "At present no adequate course in nuclear engineering can be taught at a university; the material is too secret." As a result, he asserts, too few nuclear physicists are trained each year. The few young scientists who work in AEC laboratories or who participate in AEC training programs must be contrasted with the thousands who, in Dr. Morse's opinion, would be receiving advanced nuclear physics training if that sort of training could be had in the conventional way. [37]

And it is thousands rather than a few who are needed. Robert F. Bacher, who gave up his post as an Atomic Energy Commissioner in order to become chairman of the Division of Physics, Mathematics and Astronomy at California Institute of Technology, has expressed himself as being "sure that in the days to come the limitation of trained people will be a very serious one." [38] His concern on that score is duplicated in every informed quarter. It was given fresh emphasis in the spring of 1950, when the AEC announced that the construction of the "breeder reactor" for Knolls Atomic Power Laboratory had

had to be indefinitely postponed. This was a project of very real significance. It had to be shelved because men were needed to work on hydrogen bomb problems and other immediate military matters. L. R. Hafstad, director of the Division of Reactor Development, summed up by saying, "The important point here is that the nation as a whole is short of the kind of manpower that we need in these atomic energy developments."

An increasing number of educational leaders, impressed by the difficulty of concealing a significant portion of their knowledge from the students who look to them for intellectual leadership, have simply withdrawn from contact with classified information. "I want the burden to be on Security to keep classified information away from me, rather than have the burden on me to keep scientific facts from my students," says Professor R. R. Wilson, director of the Laboratory of Nuclear Studies at Cornell University, in explaining why he declines to participate in classified work or even to look at classified documents. His sentiment has been widely echoed by others who are responsible for training youthful scientists. By divorcing themselves from all work in restricted areas, they must sometimes shun projects that are of interest to them as well as of importance to the nation. On the positive side, however, these teaching scientists can freely communicate the ideas and the information which their current inquiries may develop. In that way they avoid the building of barriers between themselves and their juniors, who, within the limits of their competence, are enabled to participate in their mentors' work. The professors' abstention from exposure to "secrets" appears to be necessary if teachers are to commune with their students, but assuredly it imposes severe and, from the point of view of the scholar, wholly irrelevant limitations upon academic work.

In the end society is the loser when the play of scientific curiosity must thus be curbed.

III

The Proper Limits of Secrecy

THE costs of secrecy are high. When the freedom of scientific exchange is curtailed, an unfavorable reaction upon further scientific development is inevitable. We pay for secrecy by slowing the rate of our scientific progress, now and in the future. This loss of momentum may conceivably be disastrous, for even from the strictly military point of view "it is just as important for us to have some new secrets to keep as it is for us to hold on to the old ones." [1] If it is unsound to suppress scientific knowledge during the long years of a cold war, the American people may one day discover that they have been crouching behind a protective wall of blueprints and formulas whose impregnability is an utter illusion.

On the other hand, no one can argue that national safety should be ignored by carefree revelation of military secrets. Surprise is an important element of a new weapon, because it reduces the likelihood of countermeasures and thus enhances the effectiveness of the development when it is first utilized. Moreover, concealment of the fact that researches are in progress may be important simply to avoid identifying the areas in which the United States does not deem itself adequately prepared. So it is plain that silence may in itself have military advantage even in connection with the more or less conventional instruments of warfare like the bazooka, the long-range bomber, and the rocket. When one's mind turns from customary military tools to the more recent engines of catastrophe such as the H-bomb, the germ invasion, and the atomic explosion, one feels even more strongly that silence may be worth its steep costs.

Can these conflicting concerns be reconciled? Is it possible to disseminate the knowledge that will lead to more knowledge, while at the same time giving respect to the military considerations just suggested?

Reconciliation is possible if an effort is made to clarify the line between scientific data and military applications of those data.

Few of the real "secrets" which this country possesses are formulas or principles beyond the grasp of others. The real secrets, chiefly, are the mechanics by which a laboratory theory is translated into a large-scale operation. The distinction is well brought out by Sir Alexander Fleming's recent reminiscence concerning the development of penicillin. While working on an entirely different problem, he chanced one day to note the extraordinary effect of a stray mould on a culture of bacteria it had contaminated. "I worked out some of the properties of Penicillin," Sir Alexander said, "and went as far as I could as a bacteriologist, but I got completely stuck because anything we did to concentrate the Penicillin which the mould produced in its culture destroyed the activity . . .

"Things remained latent from 1929 when I described Penicillin until 1939 when Florey and Chain and their colleagues set out to make a systematic study of the antibiotics which had been described. At that time I understand that they had forgotten Penicillin, but Chain, reading the literature, came across my description of it and thought something could be done chemically. They got a team together and they succeeded in concentrating the active principle about 1,000 times. This

64

concentrate they could preserve by freeze drying so they were able to accumulate a stock sufficient to test the therapeutic efficiency first on mice and then on men.

"From the first trials there was no doubt about its efficacy but then came the question of mass production during the war. The Oxford team had shown that it could be done, but this was a vastly different thing from producing it in bulk, and it was only by international co-operation of governments, scientists, industrialists, engineers and everyone down to the lowest grade workman that the production of Penicillin on a large scale was accomplished." [2]

The difficulty of translating a principle into a process has been succinctly illustrated, too, by a distinguished physical chemist. "Every boy who has had high school physics," writes Professor Frank Spedding, "knows the principles of the electric generator but this is a long way from being able to manufacture a 50,000 kilowatt generator such as is used at Niagara Falls. Here the real secret is in the technical know-how of how to produce this generator, and this secret is spread among many individuals in many professions such as miners, metallurgists, electrical engineers, chemists, physicists, etc.; no single man, if he wished, could give away the entire secret. So it is, to a much greater extent, with the so-called secret of the atomic bomb." [3]

The difference between knowledge and know-how is indeed exemplified by some of the processes which lead to the production of fissionable material in large quantities. One of the methods employed to separate U-235 from other uranium isotopes is gaseous diffusion, that is, forcing a gas against a series of metal membranes and capturing the lighter isotopes which first pass through the minute openings in these porous barriers. The understanding of the theory and mechanics of gaseous diffusion dates back to the work of, among others, Lord Rayleigh in England in 1896. But England and 1896 are a far cry from "K-25," the mile-long gaseous diffusion plant in Tennessee where uranium hexafluoride is cycled through some four thousand barriers in what is said to be the largest continuous operation under one roof any place in the world. One may doubt that Lord Rayleigh himself could have envisioned or designed "K-25."

Should we, then, seek to make a distinction between basic science and technology? Should we, in short, suppose that free trade in fundamental ideas will ensure the growth of science, while, on the other hand, guarding the details of our elaboration and effectuation of those ideas will ensure our national safety?

This differentiation is difficult to maintain systematically. The basic and therefore hypothetically innocuous science cannot readily be disentangled from the rest. As has already been observed, the forward movement of scientific achievement rarely depends upon a single flash of genius; rather, it is a consequence of the slow weaving together of many strands. Advance is built upon a selective amalgamation of the work of others, and often it is the failure or the practical limitations of one effort which suggest a fresh and finally successful approach. The realities of engineering and chemical processes frequently set the limits within which general ideas can function beneficially, with the possibility that they will stimulate still more ideas.

Nor, unless the claims of civilization are to be ignored, can the sole test of publishability of scientific work be its possible utilization in military research. Professional communication was successfully blockaded on a short-term basis during a period of active strife, and no one suffered seriously as a consequence. This does not establish that scientific freedom can or should be restrained over a span of many years. Previous pages have described the gradual and undramatic devitalization that is an inescapable concomitant of secrecy. Let us now add that abatement of scientific publication because of prolonged international tension would also entail the rigid suppression of discoveries which have great and immediate value to society in peacetime. We live, after all, in peace, not war—an uneasy peace, to be sure, and one shaken by events in Korea, but peace nevertheless. A total war may never come. All humanity prays that it will not. If mankind's intelligence is equal to the task of preserving mankind's existence, large-scale resort to arms will not occur. We must be certain that the hypothetical enhancement of martial advantage in the future is not permitted wholly to obscure the discernible enhancement of human well-being in the present.

Reference to recent developments in biological warfare research will illustrate the choice that lies open.

Since 1942, when an organization cryptically called the War Research Service began its labors, our country has actively supported investigations looking toward perfection of offensive and defensive measures for use in biological (or "bacterial" or "germ") warfare. In 1943 Camp Detrick in Maryland was set aside as the main center of work in this field, which is now under the jurisdiction of the Chemical Corps of the Department of the Army. There is no doubt about the goals of the biological warfare (BW) project, though the current operations of Camp Detrick are conducted behind an opaque wall of secrecy. "Our endeavors during the war," according to George W. Merck, the chairman of the United States Biological Warfare Committee, "provided means of defending the nation against biological warfare in terms of its presently known potentialities and explored means of retaliation which might have been used had such a course been necessary . . . Work in this field, born of the necessity of war, cannot be ignored in time of peace; and it must be continued on a sufficient scale to provide an adequate defense." To this end large sums of money and the efforts of literally thousands of persons have been devoted; their purpose has been "to extend the boundaries of knowledge concerning the use of pathogenic agents as a weapon of war and the means of protection against possible enemy use of these agents." [4]

Obviously enough, every phase of the work at Camp Detrick has military significance. If any bit of it is revealed, other nations interested in biological warfare, including potential enemies, will benefit. They will be saved time and expense in discovering infective agents and counter actions against them. They will be spared the necessity of making the same false starts that probably marked our efforts.

These circumstances, however, do not entirely offset the disadvantages of nondisclosure. The Merck Report tells us that intensive investigations were carried out at Camp Detrick into "the effectiveness of antibiotics and chemotherapeutic agents" and into "biological, physical and chemical protective measures" against "various organisms of high disease-producing power." Can we afford to keep our epidemiologists and our general practitioners unaware of the results of these activities, as we must do if our thoughts dwell exclusively on military implications? The infective agents that may be used against man in the course of BW are agents which, after all, may infect him in peacetime as well. Again, the Merck Report makes clear that extensive study was made of "biological and chemical agents which might have been used in attacking our crops," and that this "resulted in certain discoveries which will undoubtedly prove of great value to agriculture." In a dynamic economy like ours, would it be wise to ignore the "great value to agriculture" because those "certain discoveries" may also be of great value to military planners?

The intertwining of interests, the civilian and military, is nowhere more clearly apparent than in the official summary of the more important accomplishments of the Biological Warfare program up to 1946. No matter how scant may be one's knowledge of bacteriology or of the waging of war, one cannot fail to perceive that every item of the following list has potentially great significance for public health, industry, and agriculture as well as for BW:

1. Development of methods and facilities for the mass production of micro-organisms and their products;
2. Development of methods for the rapid and accurate detection of minute quantities of disease-producing agents;
3. Significant contributions to knowledge of the control of airborne disease-producing agents;
4. Production and isolation, for the first time, of a crystalline bacterial toxin, which has opened the way for the preparation of a more highly purified immunizing toxoid;
5. Development and production of an effective toxoid in sufficient quantities to protect large scale operations should this be necessary;
6. Significant contributions to knowledge concerning the development of immunity in human beings and animals against certain infectious diseases;
7. Important advances in the treatment of certain diseases of human beings and animals, and in the development of effective protective clothing and equipment;
8. Development of laboratory animal propagation and maintenance of facilities to supply the tremendous number of approved strains of experimental animals required for investigation;
9. Application of special photographic techniques to the study of airborne micro-organisms and the safety of laboratory procedures;
10. Information on the effects of more than 1,000 different chemical agents on living plants;

69

11. Studies of the production and control of certain diseases of plants.[5]

Since January 1946 about 160 papers and monographs embodying BW researches have been published. The fact that these materials are available for general use reflects enlightened awareness by the Army that the science of peace and the science of war have many common interests. Dr. Rosebury in his excellent book, *Peace or Pestilence,* has traced the value of these reports for "healthy science." [6] Camp Detrick studies on synthetic plant-growth regulators have provided tools to aid in basic research into "the nucleus which dominates the activities of the living cells." The crystallization of botulinus toxin, an unprecedented accomplishment, is likely to spur the final isolation of other bacterial toxins and has "put in the hands of the chemist powerful tools for exploring some of the basic problems of disease." Study of viruses that produce animal diseases has yielded new methods for recognizing them promptly as well as effective vaccines for protection against them.[7] The steps taken at Camp Detrick to control accidental airborne infections "have proved valuable not only in research with highly infective agents there and elsewhere but also in work that requires the exclusion of germs, as in the commercial production of biologicals like liver extracts, which must be handled in a germ-free environment because they are damaged by any attempt to sterilize them with heat or chemicals." The BW experiments on infection carried through the air "have also made available exact methods and refined techniques to attack the most important group of human diseases still uncontrolled by sanitation—the respiratory infections, like influenza and tuberculosis."

The catalog of positive advances made possible by biological warfare research is far from exhausted by these instances, which in any event deal only with immediately foreseeable benefits. These discoveries, like other fundamental data, are likely to be of yet further help in ways which no one can as yet know. The given instances sufficiently suggest, however, that a publication policy that adhered strictly to a "guns instead of butter" philosophy would have deprived the nation of a very considerable amount of butter, indeed. The social costs of secrecy are readily seen here, just as they would be if the researches of our agronomists and animal husbandrymen were to be "classified" for fear that a potential enemy might use them to increase its food resources.

When one turns to industrial applications of military researches, the choices become less plain because they are not colored by moral convictions regarding human health. Even so, there is cause for concern in the fact that American industrial efficiency has not been given as much consideration as perhaps it deserves. For instance, there has not yet been full publication of the information gained by the National Defense Research Committee concerning the behavior of materials under strain and pressure. Fundamental knowledge acquired through studies of the various reactions occurring when a gun is fired would have significance for high-compression technology generally. During the war American scientists developed a machinable metal, "Alloy X," which possesses remarkably high strength, moderate ductility, and hot-hardness and is thought to be capable of numerous future applications. But because the erosion-resistant qualities of Alloy X make it useful for lining the barrels of high velocity guns, even the basic metal from which it was evolved is still a secret withheld from American metallurgists.[8] If military purposes are thought to be advanced by suppressing knowledge of these sorts of scientific finds, we should at least be aware that suppression does not contribute to an ever more abundant economy.

If, then, a general proposition may be suggested, it is this:

71

Secrecy ought not to be readily attached to scientific or technological matters merely because in some aspects they have military significance. It should be attached unhesitatingly if their sole significance is a military one. Application of this differentiation may be clarified by referring to the several fields of earlier discussion.

Much Biological Warfare research, for example, has been released, but, notwithstanding a generally liberal publication policy, a great deal remains steeped in secrecy. Despite the Merck Report's assurance that Camp Detrick developed "methods for the rapid and accurate detection of minute quantities of disease-producing agents," no details concerning those methods have yet been reported. This seems an indefensible exaltation of military values over human needs. On the other hand, suppression of reports concerning the containers developed for disseminating infectives seems entirely justified. There is no discernible civilian need for specially constructed devices for spreading pathogenic agents, which have been aptly characterized as "BW munitions." They constitute part of the secret techniques of war rather than part of the life-enriching treasury of science. If they remain secret and unrevealed forever, mankind will be the gainer rather than the loser.

The Atomic Energy Commission has refused to declassify a research report on the effects of exposure to a certain chemical compound, because the report was written at Los Alamos and the inference might therefore be drawn that the chemical in question is used in connection with bomb manufacture. Similarly, at the Argonne National Laboratory a report of experiments on the properties of certain uranyl salts was placed under restriction, apparently because the experimenters had suggested that their study might possibly shed light on the separation of uranium isotopes as well as other chemical processes. Suppression of these types of knowledge seems of doubtful wisdom. They may be importantly useful in the planning of industrial operations or in the conduct of researches wholly unrelated to the production of bombs. By way of contrast, chemical research during the war made possible the perfection of unorthodox hand devices and techniques of sabotage for use by guerrilla and resistance forces, looking toward maximum destruction of enemy personnel and property. Most of the weapons produced for this purpose were simple in design and were chiefly of an explosive or incendiary nature. Unfortunately the unconventional devices that were created for field use during the war are suitable for employment by lawless, terroristic, or subversive elements in time of peace as well. The knowledge embodied in these weapons is so unlikely to have legitimate application that continued restrictions upon its publication are fully warranted.[9]

A distinction must properly be drawn between, say, information concerning neutron cross sections of the heavy metals (which, being valuable for further physical research, ought to be revealed) and information concerning the design or mechanism that prevents premature disintegration of an atomic bomb before it has efficiently utilized its charge of fissionable material (which, being essentially a military device, may properly be concealed). A distinction must be drawn between, on the one hand, a new understanding of aerodynamics and, on the other, the plans of a specific military aircraft that undertakes to utilize the new understanding. In short, the design of weapons, reports about their performance and properties, the design of large-scale plants for their production, and, occasionally, specific instruments or processes can be kept under flexible restrictions without any very likely effect upon industrial or scientific advance. But care must be exercised to avoid confusing these matters with principles and practices which expand the edges of understanding and which may be pieced together with other bits of knowledge for the well-being of mankind. While it is true that the latter may conceivably benefit a potential enemy in some particular, the risk of that benefit is more bearable than the sapping of our own strength.

It would be unfair to suggest that this commonsensical conclusion has been beyond the grasp of our nation's military and atomic energy authorities. Quite to the contrary, there is every reason to believe that existing basic policies are not inharmonious with it.

Unfortunately, however, the effectuation of those policies has been retarded by two forces. One is the force of official inertia, the reluctance to exercise judgment incisively and boldly, the unwillingness to accept responsibility for disclosing information which a later critic may maintain should have remained secret.[10] The other is the force of a badly misled public opinion.

Enlightenment of popular sentiment is difficult so long as political leaders violently denounce the imparting of knowledge as though it were a plot to advance the fortunes of Soviet Russia. Both the Atomic Energy Commission and the services have occasionally manifested readiness to lower the barriers which decelerate scientific progress and which block public understanding of giant governmental efforts to enlarge our resources.[11] Their inclinations in this respect are not stimulated by criticisms such as those addressed to the AEC by a distinguished Senator, who deplored the AEC's reproduction of a photograph of the outside of a small model of the Brookhaven proton-synchrotron ("or some such thing"), a nonsecret research tool, or a prominent Representative's perturbation that the AEC had revealed that Brookhaven has a 420-foot tall tower that emits smoke "which can sometimes be seen for miles around." [12] If secrecy is permitted to become a fetish, rational judgments lose their relevance.

The hope for science in this country and for the nation's

security is that the public at large will shed its fears, grow in understanding, and cease credulously accepting assertions that safety lies in secrecy. Secrecy is antithetical to the spirit of science.[13] It is socially hurtful. Only for brief periods can it be practiced without destroying the scientific superiority it is intended to preserve. Today the United States holds a position of dominance in world science largely because of its rich resources of technical and scientific manpower, coupled with material facilities that cannot be duplicated by the impoverished countries of Europe and the Orient. Unless this country dissipates its advantages by artificially limiting what the rising generation of scientists may be permitted to learn, its strong ranks of talented, well-trained humans rather than its possession of a body of knowledge are probably the chief guarantor of America's future leadership.

IV

The Standards and Mechanics

of Security Clearance

SO LONG as war is thought to be just around the corner, every great nation devotes a large part of its wealth and ingenuity to efficient military preparations. In so far as those preparations may involve the development of weapons or equipment, the United States, like other countries, will seek to conceal progress from the eyes of potential enemies in order to maintain the advantage that inventive skill may temporarily give it. Moreover, since the element of surprise is itself deemed a military asset, not only the details of mechanisms but also the extent of their availability may sometimes be regarded as "military secrets," to be withheld from the knowledge of competitors if possible.

Today, as earlier discussion has emphasized, the scientist is the nation's armorer to an extent never before approximated. He is himself the creator rather than merely the guardian of military secrets. Some part of his information must be available to all if civilization is to progress. Other bits of his knowledge may justifiably be buried for short-range military reasons. The dividing line is not hard and fast. The tug and pull of competing considerations will influence the pattern. Somehow, nevertheless, a pattern will emerge. The line is drawn, uncertain in direction, fluctuating in purpose, and unstable in duration though it may be. Once it has been drawn, it momentarily determines the dimensions of the area of secrecy. And once that area has been defined by appropriate public authority, there immediately arises a proper interest in assuring that all who work within it will scrupulously observe its boundaries. So long as the boundaries exist, they must not be ignored.

Obedience to public commands is conventionally compelled by penalties upon the disobedient. The fear of detection and punishment deters transgressions. But, as daily sensations remind us, the threat of retribution does not wholly eliminate criminal or other antisocial conduct. At best, misbehavior is merely somewhat diminished. Hence society appropriately seeks for other measures, and especially measures of a preventive character, to forestall injuries to it. In the context of the present discussion, the measure chiefly relied on as a preventive of unreliability within the zone of secrecy is the personnel security program. Through this program the government hopes to sift out the persons who, like the faithless English scientists Alan Nunn May and Klaus Fuchs, might flout restraints which national military needs have generated. Excluding potential malefactors from the area of secrecy may be a surer shield than would be the most severe punishment of wrongdoing after the event. Since the world includes persons who are undisciplined or corrupt or treacherous, there is wisdom in trying to identify them before they are permitted to deal with matters of immediately vital public safety.

The prime purpose of the personnel security program is to assure that acts of sabotage will not occur and that "secret information" will not pass into the hands of others than those to whom it has been entrusted. Thus justified, the program extends to many types of personnel besides scientific workers. The construction gangs that erect the specially designed build-

77

ings of an atomic energy installation, for example, must be "cleared," as must the maintenance employees, the clerical staff, the guards, and all the others whose jobs involve physical access to restricted areas or use of "restricted data." Similarly the businessmen who wish to bid on contracts to supply certain types of military equipment must be "cleared" before they may read the specifications that will shape their bids; and when a contract is awarded, the process engineers and other technicians, as well as many production workers who are involved in executing it, must be investigated and their "security" established to the satisfaction of public authorities.

A significant qualitative difference does, however, set apart the security investigations of scientists.

In the generality of cases affecting nonprofessional employees the investigators are chiefly concerned with the character of the individual under investigation. Does his past record suggest irresponsibility or inattention to regulations governing his employment? Is he a drunkard who might carelessly reveal confidences? Is he constantly in debt and therefore perhaps susceptible to bribery? Is his an abnormal personality? Does he have a serious criminal record that indicates habitual disregard of obligations to society?

The cases in which a scientist's security has been questioned are in marked contrast. In scarcely a single case involving a scientist, so far as diligent inquiry has disclosed, have the issues been of this sort. The scientists' cases have involved not character, but attitudes; not behavior, but associations; not personality, but opinion.

"Reliability" in these respects is chiefly the concern of the Atomic Energy Commission and the military services. The scope of their authority and the procedures they employ warrant consideration.

78

Personnel Security in the Atomic Energy Commission

The Atomic Energy Act of 1946 emphasizes in many of its sections the policy of hoarding our real or supposed "atomic secrets." As a specific safeguard against revealing these treasures to individuals who might be unworthy of trust, the Act provides that—

1. No individual employed by a contractor or licensee having relations with the AEC may be permitted by his employer "to have access to restricted data until the Federal Bureau of Investigation shall have made an investigation and report to the Commission on the character, associations and loyalty of such individual and the Commission shall have determined that permitting such person to have access to restricted data will not endanger the common defense or security"; and

2. With exceptions not now material, "no individual shall be employed by the Commission until the Federal Bureau of Investigation shall have made an investigation and report to the Commission on the character, associations, and loyalty of such individual."

Thus the Atomic Energy Commission is empowered and directed to pass on the eligibility of all who find "restricted data" essential to performance of their scientific duties. In the main these are not scientists who are themselves a part of the AEC staff. The AEC directly employs no more than a hundred scientists, chiefly in administrative rather than research activities. The scientific work that interests the AEC goes forward in university or industrial laboratories or in huge installations that are owned by the AEC but operated by a contractor. The Carbide and Carbon Chemicals Corporation, for example, administers the gigantic plants and laboratories in Oak Ridge, and most of the scientists who work there are its employees, not the Government's. Similarly, the weapons research carried on by scientists at Los Alamos is one of the contractual responsibilities of the University of California, which also operates the Radiation Laboratory in Berkeley. So with each of the major centers of work in the field of atomic energy; the laboratories may have been created by the United States, but they are administered by academic or industrial contractors, which hire their own scientific staffs, subject always to the Commission's granting "security clearance" that will permit access to restricted data. When we speak of atomic energy scientists, therefore, we refer for the most part to the faculties of numerous educational institutions; or to the employees of such concerns as Monsanto Chemical Company, the operator of an AEC laboratory at Miamisburg, Ohio, where highly classified process, research, and development work is carried out; or to the staffs of installations like the Argonne National Laboratory, which is operated by the University of Chicago as chief contractor aided by a council of thirty other institutions. The various possible extensions of the program into private employment are readily suggested by the names of the corporations which, being interested in industrial applications of nuclear energy, support the University of Chicago's basic atomic and metals research:

Aluminum Company of America, American Tobacco Company, Beech-Nut Packing Company, Bethlehem Steel Company, Celanese Corporation of America, Commonwealth Edison Company, Copper & Brass Research Association, Crane Company, E. I. du Pont de Nemours & Company, Fairchild Engine & Airplane Corporation, Inland Steel Company, International Harvester Company, Pittsburgh Plate Glass Company, Procter & Gamble Company, Reynolds Metals Company, Shell Development Company,

Standard Oil Company (Indiana), Standard Oil Development Company, Sun Oil Company, United States Steel Corporation, and Westinghouse Electric Corporation.

The AEC's duty to consider the reliability of many thousands of individual employees has not been an easy one. Between January 1, 1947, and April 28, 1949, the Commission occupied itself with security matters at 151 of its 262 formal meetings, and spent perhaps a third of its entire meeting time on personnel security matters alone.[1]

The Commission inherited a large operation from the Army, which had administered the atomic energy program during its fast-growing infancy. The Army's security procedures had been, to put the matter as mildly as possible, somewhat primitive. Investigations of all employees were made under the direction of Military Intelligence. Those who were suspect were rather abruptly ejected. A man who was subject to being called into military service might find himself hurriedly summoned to leave his scientific researches and to enter forthwith upon less onerous duties in some military outpost. Those who could not be transferred in this way were simply dismissed summarily. Some of the quick decisions in that period were no doubt sound. Some probably were not. There simply was no time to be sure which was which, and war always causes casualties.

In most cases, of course, security clearance was not denied the scientists who were equipped to participate in the program. After all, a large organization was needed, and needed urgently. If every doubt were resolved against every employee, too many might have been ushered out; and there was no time in which effective replacements could be trained. As General William J. Donovan, the wartime head of the Office of Strategic Services, once remarked, "You can have an organization that is so secure it does nothing," or you can decide to move

81

forward by taking some chances. "If you're afraid of wolves," he added, "you have to stay out of the forest." [2] By and large the Army's Manhattan Engineer District was not afraid of wolves. It granted clearances.

When the AEC took over the MED's operations and its staffs, however, Congress directed that all who remained in work involving access to restricted data must be reinvestigated. To be sure, they were given interim clearance; but continuation of their employment rested on the AEC's finding, after a fresh investigation by the Federal Bureau of Investigation, that the common defense or security would not be jeopardized by their presence on the projects. To this large number of persons who were to be reinvestigated and reappraised were added the thousands of new recruits who entered the rapidly expanding atomic energy field after the war. From January 1, 1947, to April 30, 1949, a total of 141,469 individuals were evaluated by the Atomic Energy Commission. During 1949 there arose a total of 37,561 new personnel clearance cases, and this number may sharply increase as new installations come into being.

Obviously, a fairly elaborate administrative machine is needed to cope with a case load of these dimensions. The investigations themselves are not a burden to the Atomic Energy Commission, because they are conducted in each instance by agents of the Federal Bureau of Investigation. Contrary to a widely prevalent belief, however, the duty of evaluating the investigation reports rests on the AEC rather than the FBI, which is wholly without responsibility for reaching conclusions as to the significance of the facts and rumors its inquiries have revealed. The actual mechanics of decision are these:

1. The FBI report is first considered in a subunit of one of the AEC operations offices, which are located in Chicago, Hanford, New York, Oak Ridge, Santa Fe, Schenectady, and Arco, Idaho. Each of these offices has primary responsibility

for certain of the activities and installations of the Commission. The initial review of the investigation report is undertaken by the office that has operational jurisdiction over the particular enterprise in which the affected individual will do his work. If, for example, a physicist were recruited for the Knolls Atomic Power Laboratory, administered by the General Electric Company in Schenectady, New York, his security clearance would first be considered by the staff of the manager of the operations office in that city; if he were to work in the atomic energy project of the University of Rochester, where research is done under contracts initiated by more than one AEC office, the papers would "follow the contract" and would accordingly go to the New York Operations Office or to Oak Ridge as the case might be; and if he were to be employed by the Monsanto Chemical Company in the Miamisburg laboratory in Ohio, the file would be studied by AEC officers under Oak Ridge direction.

2. At this stage the investigation reports are analyzed by members of the local security staff and, in difficult instances, by others, including legal counsel and the manager himself. If the analysts decide that "employing such persons or permitting them to have access to restricted data will not endanger the common defense or security," the manager (or his delegate) may grant the desired security clearance. Whenever doubts remain, and especially where certain particular types of evidence appear in the record, the whole file must be forwarded to the AEC's Division of Security in Washington. Despite indications to the contrary in some of the Commission's publications, the fact is that clearance may not be denied by the local Manager of Operations, though of course he may recommend that it should be withheld. In other words, the power to grant clearances has been largely decentralized, but the power to deny clearances has thus far been reserved in a central staff agency. The doubtful cases are considered at head-

quarters by several strata of reviewers in the Division of Security, and the authoritative decision as to granting or withholding clearance is made there.

When clearance is denied, provision is made for subsequent review procedures, which may be briefly summarized as follows: If the affected individual is already employed under a prior clearance giving him access to restricted data, he receives notification that his clearance is about to be withdrawn. If he so desires, he may have a hearing before a "local personnel security board," appointed by the Manager of Directed Operations for the area in which he is employed. In most cases involving scientists, hearing boards thus far appointed have been composed of a member of the AEC administrative staff, an attorney of reputation in the locality, and a scientist who has insight into the relationship of the individual to the project as a whole. The local board so constituted makes a recommended decision to the local manager, who in turn forwards his recommendation to the Commission's General Manager in Washington. If this is adverse to the employee, he may request further consideration of the case by the Personnel Security Review Board, which may also be asked by the General Manager, on his own initiative, to review any case upon which he desires further advice. The Personnel Security Review Board has no power of final decision; its action is a recommendation to the General Manager to assist him in his final determination as to security clearance. The General Manager may, of course, present important policy considerations to the Commission itself.

The Commission has taken great care to appoint an advisory body that would command public confidence. The original Personnel Security Review Board consisted of Owen J. Roberts, former Supreme Court justice and now dean of the University of Pennsylvania Law School; Karl T. Compton, then the president of the Massachusetts Institute of Technology; Joseph C. Grew, former Undersecretary of State; George M. Humphrey, president of the M. A. Hanna Company; and H. W. Prentis, Jr., president of the Armstrong Cork Company. According to its minutes, that board met on seven occasions between July 1, 1947 and September 4, 1948.[3] During this period some forty recommendations were made to the General Manager. The initial members of the Personnel Security Review Board tendered their resignation en masse during the summer of 1948, and ceased functioning in September of that year. They were not replaced until March 10, 1949, when the AEC announced a "permanent Personnel Security Review Board" consisting of Charles Fahy, a Washington attorney and a former Solicitor General of the United States who had had broad governmental experience; Arthur S. Flemming, a former United States Civil Service Commissioner who is now president of Ohio Wesleyan University; and Bruce D. Smith, director of the United Corporation and formerly an official in the War Manpower Commission. More recently Mr. Fahy was appointed a judge of the federal Court of Appeals in the District of Columbia, his place on the review board being taken by Ganson Purcell, who practices law in Washington after having served as chairman of the Securities and Exchange Commission.

This impressive machinery for review and for possible modification of previous decisions works, however, only in the cases that involve "old hands," the people who have been cleared previously by the Manhattan Engineer District or by the AEC itself and who are still at work. Those who seek clearance now in order to commence scientific labors requiring access to classified data have no regularized means of obtaining review of an adverse determination. In their cases, if clearance is withheld, the matter is ended. No charges are stated, no hearing is held, no appeal is possible. Clearance is denied. At present this total absence of any formalized device to avoid unsound decisions affects many more people than does the presence of the Personnel Security Review Board. The failure to safeguard the rights of applicants for clearance is one of the most serious shortcomings of the AEC. The criticisms to which this deficiency gives rise are discussed more extensively in a later chapter.

From the very beginning of decentralization the AEC instructed its representatives in the field to make favorable decisions only in cases in which no "substantially derogatory information" had been brought to light concerning the applicant. "Substantially derogatory information" was but sketchily defined in the instructions which the AEC's staff received.[4] Not until January 5, 1949, was the Commission able to formulate and announce the factors that may create serious doubts concerning eligibility for clearance. On that day it published its "Criteria for Determining Eligibility for Personnel Security Clearance." [5]

Even these declared criteria are merely suggestive rather than definitive, for the Commission recognizes that no formula can embrace all the variants of human personality and organizational needs. Thus, for example, information that would probably raise doubts about the character of an unknown job seeker might be deemed wholly insignificant in the case of a man who had rendered long and satisfactory service in an atomic energy installation under the close observation of responsible supervisors. Moreover, as the Commission puts it, "a determination must be reached which gives due recognition to the favorable as well as unfavorable information concerning the individual *and which balances the cost to the program of not having his services against any possible risks involved.*" This is a point of especially great importance in connection with the clearance of mature scientists. The number of trained persons is inadequate to supply the nation's present needs. If a scintilla of doubt about a man's reliability

86

were to lead automatically to rejection of his potential contributions, we might indeed find ourselves with "an organization that is so secure it does nothing." For this reason the Commission's criteria have been set forth not as decisional principles, but as determinants of the categories of "derogatory information" which create serious doubts. The criteria do not foreclose the possibility that those doubts may be dissipated by other information; they merely serve to identify the cases which call for close attention.

"Category (A)," as set forth by the Commission, embodies types of derogatory information which, standing quite alone, establish a presumption of security risk. In any case of this sort the local manager has no power to resolve doubts in favor of clearance; the file must be at once forwarded to the Division of Security. The topics which Category (A) touches upon include information that the individual or his spouse has engaged in activities involving sabotage, espionage, treason, or sedition, or has had relations with foreign spies or "representatives of foreign nations whose interests may be inimical to the interests of the United States." So far as can be ascertained, information of this sort has been developed in very few if any of the nearly 200,000 cases upon which the Commission has now passed. Category (A) also includes:

1. Continued membership in an organization after the Attorney General has declared it to be subversive, or prior activities in a capacity which should have made the individual aware of its subversive purposes;
2. Advocacy of violent revolution;
3. Omission from or falsification of a Personnel Security Questionnaire or Personal History Statement;
4. Serious disregard of security regulations on former occasions;
5. Insanity;

87

6. Conviction of felonies indicating habitual criminal tendencies; and

7. Addiction to excessive use of alcohol or drugs.

Apparently these types of seriously derogatory information are rarely disclosed by FBI investigation of persons seeking AEC employment. Members of the AEC staff at several locations have asserted that they know of no case of this sort in which a scientist has been involved. But since it has been impossible to obtain a central office confirmation of these field officers' impressions, one cannot say flatly that there never has been a Category (A) case. What can be confidently asserted, however, is that almost all the cases which have required thought before a decision was reached—and the total number of these is only slightly above two thousand—have arisen under "Category (B)." [6]

Category (B) like Category (A) lists matters that the Commission says would ordinarily warrant a denial of clearance. In these cases, however, the Manager of Operations is empowered to grant clearance if, on the whole record, he thinks it proper; he may recommend against clearance if he is convinced that the presumption of risk has not been overborne by other evidence; or in borderline cases he may pass the buck to the Director of Security in Washington without expressing a judgment one way or the other.

Category (B) cases include those in which either the individual or his spouse—

1. Has shown "sympathetic interest in totalitarian, fascist, communist, or other subversive ideologies";

2. Has been sympathetically associated with any members of the Communist Party or with "leading members" of any other organization the Attorney General has declared to be subversive;

3. Has been identified with a "front" organization when

88

the individual's personal views "are sympathetic to or coincide with subversive 'lines' ";

4. Has been identified as a part of or sympathetic to a group of subversives who are infiltrating a nonsubversive organization;

5. Has close relatives who live in countries which might exert pressures upon them as a means of forcing the individual to reveal sensitive information or commit sabotage;

6. Lives at the same premises or visits or frequently communicates with friends, relatives, or other persons who have subversive interests and associations;

7. Has formerly had close association with such friends, relatives, or others, now interrupted by distance but perhaps likely to be renewed in the future;

8. Has conscientious objection to military service when the objection is not clearly a product of religious conviction;

9. Has tendencies demonstrating inability to keep important matters confidential; carelessness in observing regulations concerning the use of restricted data; dishonesty; or homosexuality.

The chief differentiation between Category (A) and Category (B) is easy to see. Almost all the situations that fall in Category (A) are matters of personal conduct or character. Almost all the situations that fall in Category (B) are matters in the realm of ideas or associations which do not reveal any actual misconduct on the part of the individual.

This illuminates and emphasizes what is frequently overlooked in descriptions of the personnel security system. The finding that underlies a decision to withhold clearance need not be that the individual has been wicked or, even, that he probably will be wicked. All that is needed is a finding that the individual *may* be disposed to be wicked or careless at some indeterminately future time. In truth, all that is needed is a finding that the individual's spouse might designedly or otherwise acquire from him and subsequently transmit to others information that has not been released to the public at large.

Predictive or, if you will, precautionary findings of this sort involve very different mental processes from those that occur in the ordinary trial of an issue of fact. In most conventional fact-finding proceedings, an effort is made to achieve an evidential reconstruction of an event that has already occurred. In security proceedings the effort is, instead, to formulate a judgment about the degree of possibility that an event will occur in the future. The extent of the risk that a particular individual will be faithless is not subject to conclusive demonstration. A judgment concerning it involves hypotheses, impressions, experiences, and generalized prejudices (favorable or unfavorable to the applicant), which are brought to bear consciously or, often, unconsciously. It must be clear, therefore, that what is really being appraised in a personnel security case is not any particular question of fact but is, in a word, a man.

Nowhere is this more specifically recognized than in the AEC's "Memorandum of Decision Regarding Dr. F. P. Graham, December 18, 1948." Dr. Graham, later a United States Senator from North Carolina, was at that time president of the University of North Carolina. He was also the president of the Oak Ridge Institute of Nuclear Studies, a nonprofit organization of twenty-four southern universities established to assure broad regional participation in the atomic energy educational and training activities that center at Oak Ridge. To give the Institute effective guidance in its development Dr. Graham might occasionally require access to restricted information, and so he had to be "cleared." The FBI report on Dr. Graham that was laid before the AEC showed that he had "been associated at times with individuals or organizations influenced by motives or views of Communist derivation." Should clearance therefore be denied? The AEC in one of the two written opinions about personnel security that it has allowed to become public held that clearance should issue. " 'Associations' of course have a probative value in determining whether an individual is a good or bad security risk. But," concluded all five members of the Commission, *"it must be recognized that it is the man himself the Commission is actually concerned with,* that the associations are only evidentiary, and that common sense must be exercised in judging their significance. It does not appear that Dr. Graham ever associated with any such individuals or associations for improper purposes; on the contrary, the specific purposes for which he had these associations were in keeping with American traditions and principles. Moreover, from the entire record it is clear in Dr. Graham's case that such associations have neither impaired his integrity and independence, nor aroused in him the slightest sympathy for Communism or other anti-democratic or subversive doctrines."

So Dr. Graham was tried as a man, was found to be worthy of trust, and was cleared in order that the country might have the advantage of his continued participation in the atomic energy program.

Senator Graham, of course, is not typical of the men who may be involved in a security case. He was well known. His actions over many years were publicly recorded. The purposes of his associations were readily inferable from the course of his conduct in other connections. How can the Commission concern itself with "the man" instead of "the associations" in cases where the individual is less prominent and his motives less obvious? The procedures by which this is sought to be done will be examined in a later portion of this discussion.

Personnel Security in the Military Services

Important though it is, the personnel security work of the AEC does not touch so many scientists and technologists as does the security program of the armed services. The latter, which may be called military clearance in order to distinguish it from the AEC processes just considered, applies to three large and wholly separate groups of scientific personnel.

In the first group are more than 12,000 scientists employed directly by the Army, the Navy, or the Air Force for work in installations like the Edgewood Arsenal, the Aeroballistics Facility, the Navy Electronics Laboratory, or the Alamogordo guided missiles project. The number of 12,000 includes only civilians with professional civil-service ratings as physical, biological, or agricultural scientists and thus excludes all military personnel who may also be assigned to scientific work.

In the second group are government scientists employed by civilian agencies but engaged in research on military projects. The National Bureau of Standards, a unit of the Department of Commerce, has, for example, undertaken for the Navy a study of the aerodynamic characteristics of aircraft bombs, finned projectiles, and rockets. Similarly the Bureau of Mines of the Department of the Interior has conducted on behalf of the Air Force an investigation of aviation fuels which might influence design of new engines and equipment. And the Tennessee Valley Authority, more or less as a by-product of its research on phosphatic, nitrogenous, and potassic fertilizers, has explored the adaptation of chemical products and processes to the manufacture of munitions. When projects of these sorts involve secret material, all those who may have access to the research data must be cleared even though they are the employees of other official branches of the Government. If the responsible military department withholds clearance of one of these federal scientists, he must simply be assigned to other work.

The third group, no doubt larger than the other two combined, comprises the scientific personnel employed by educational and other nonprofit institutions or by industrial corporations that have contracted to do classified work for one of the military agencies.

Because so much of the nation's developmental research and productive enterprise is linked with the making or improvement of military articles, the grasp of military clearance has extended far beyond the conventional boundaries of government into the realm of purely private employment. It is most important to note that the procurement agencies of the armed forces have exclusive and discretionary power to determine the extent to which work on contracts is to be classified. Since the military orders of our own government and our political allies absorb an ever-increasing share of American industry, a very large segment of all employment must quickly become subject to personnel security procedures unless the authority to impose classification restrictions is moderately exercised. Without reference to questions of organization or procedure, this prospect can but alarm all who value the American tradition of civilian freedom from military surveillance and restraint. The tendency to "overclassify" may have bitter consequences if not rigorously curbed.

Matters of principle aside, overclassification slows down vital production; when more and more persons must be cleared before work can be commenced, the end result is inefficiency. According to a dispatch by Walter H. Waggoner to the *New York Times* on June 19, 1949, "Officials estimate that as many as 20,000 to 50,000 technicians, engineers, scientists and other key industrial employees either are not working or have only interim clearance on their jobs pending their

specific approval for handling classified processes or materials.
. . . The mounting accumulation of security investigations
to be made of industrial workers threatens not only to be a
drag on important defense contracts that should be completed
promptly, officials believe, but also to be a staggering administrative task for the National Military Establishment."
Prominently listed among remedies that were being considered to reduce "the welter of investigations clogging the Government's security offices" were declassification of many processes and products and lowering the classification on others
so that fewer persons would require clearance.

Scientists Employed by the Military

The Secretary of any one of the three military departments
may remove any departmental employee whose dismissal he
regards as "warranted by the demands of national security."
This power, conferred by a statute that was enacted in 1942
"To expedite the prosecution of war," is summary and uncontrolled.[7] The only procedural nicety the law prescribes is
that "within thirty days after such removal any such person
shall have an opportunity personally to appear before the
official designated by the Secretary concerned and be fully
informed of the reasons for such removal"; then he may sub-
mit "such statements or affidavits, or both, as he may desire
to show why he should be retained and not removed."

As might be expected, this abrupt authority has been exercised brusquely on a number of occasions. It is to the
credit of the armed services that they have sought to moderate
their procedures. They have seriously attempted to avoid
judgments that "demands of national security" require the
degradation of professional men whose chief offense is non-
conformism. Moderation and restraint are still needed.

Scientists who are employed by one of the military departments are, like all other federal employees, subject to

94

removal under the terms of the "Loyalty Order," which will be considered in later pages. But "loyalty" and "security" may not be coextensive. If a man is thought to be disloyal, of course he is a "poor security risk." On the other hand, a man may be adjudged entirely loyal to his country and yet be deemed objectionable from the standpoint of security because he drinks excessively or his wife holds unorthodox opinions. The distinctions between, as well as the overlapping of, security and loyalty have caused organizational difficulties for the services that each of the three has attempted to surmount in a different way.

The Army's civilian employees, like all government personnel in civilian agencies, are first investigated by the Federal Bureau of Investigation. A supplemental investigation is also made by Military Intelligence. The initial determination that dismissal is warranted by security considerations is then made on the basis of the investigators' report supplemented by any material that may be available in intelligence files. In form the determination is a decentralized one, for the first decisive step is taken by the commanding officer of the area, advised by intelligence officers. He may suspend an employee up to ninety days, at the end of which period he must either reinstate the affected individual or recommend to the Secretary of the Army that he dismiss the man. This recommendation is in due time reviewed by the Intelligence Division, which passes it along to the Secretary's office with a statement of its findings and proposals. If the Secretary's Personnel Division agrees that charges should be pressed, they are sent to the employer with a letter of removal, which takes immediate effect.

There is always reason to fear too great a readiness to act adversely on very slight provocation in cases which involve unpopular elements and in which no opportunity is afforded to hear the other side of the story. Men who are "trigger happy" are unlikely to decide wisely in matters often marked by deli-

95

cacy of nuance. Recognizing this, the Secretary of the Army in 1948 created a Security Review Board with a civilian chairman, to act as his adviser in these matters. Persons who have been summarily dismissed are afforded an opportunity to appeal to this body. Since every "loyalty case" may also be deemed a "security case" in a department that has authority to dismiss any employee summarily if security is involved, the Army does not observe the procedural and organizational aspects of the President's Loyalty Order; instead it proceeds in each instance under the powers conferred by Public Law 808, the 1942 summary removal statute. The Army's Security Review Board sits only in Washington, and is often only theoretically accessible to those who most urgently desire to appear before it. No funds are provided to make possible the attendance of the affected individual or his witnesses, so that many cases must be reviewed on the basis of documents and written protestations of innocence rather than on the basis of living evidence and arguments. Even so, the Security Review Board recommends to the Secretary that he set aside the decisions in about twenty per cent of all security dismissals, and in a still higher percentage of the cases that are appealed to it.

The Navy Department is even more summary in its acts under Public Law 808. If an employee "occupies a key position or a position of trust" (as many of the Navy's scientific personnel do), he may be removed on security grounds without any hearings whatsoever, whether in Washington or elsewhere, before or after the event. The employee receives a brief explanation of the reason for his having been ousted. Then, if he chooses, he may file with the Secretary a protest against the action. That is all, in theory. As a matter of fact, however, the Navy goes a good deal beyond this in providing procedures which, at least on the surface, are fairer and more orderly. Where the evidence raises a question about an employee's loyalty, a hearing is provided in the field, with an opportunity to seek review by the Navy Department Loyalty Appeal Board. An adverse determination by that body is subject to an appeal to the Civil Service Commission's Loyalty Review Board. Of course even if this board recommends the exoneration of the employee, the Secretary of the Navy still retains power to terminate the employment on the ground that "security" so demands. It is perhaps pertinent to note that the chairman of the Navy's appeal board, concerned with both "security" and "loyalty" cases, is the Department's director of personnel, who is generally regarded as the author of the law by which summary removal has been made possible.

The Department of the Air Force operates still differently. Acknowledging that there is a probable though not inevitable nexus between loyalty and security, it provides a single procedure for both types of cases. If a man is dismissed because the commanding officer deems him to be either a security risk or a disloyal person, he may ask for a hearing before a Loyalty-Security Hearing Board. The hearing boards are decentralized, thus overcoming the geographical difficulty that impairs the utility of the Army's Security Review Board. But since the Loyalty-Security Hearing Board is drawn from local personnel, perhaps dominated in some instances by the tradition of subservience to the commanding officer whose judgment is formally under review, the blessing may not be altogether unmixed. It is noteworthy, however, that Air Force regulations require a majority of the hearing board's members to be chosen from civilian rather than military personnel.

If the local board's decision is adverse, there is in any event an opportunity for appeal to the Air Force Loyalty-Security Review Board, which sits centrally and is not affected by the same psychological pressures that may conceivably operate locally. Where the charge involves loyalty, there is yet another appeal, this time to the Civil Service Commission Loyalty Review Board. Let it be emphasized, though, that the Air Force like the Navy is not bound by a favorable judgment of the Loyalty Review Board. In the end, of course, a man may still be discharged because he is thought to be a security risk, even if the highest authority in the land were to adjudge him "loyal." Still, the Air Force does seem to go farther than the other two services in waiving the discretionary summariness with which Congress has endowed it.

Research and development programs are heavily relied upon by all three of the armed services as vital adjuncts to forces in being. As a report to the Army's General Staff forthrightly declared, "Success in any future war will depend as much on the effective use of all the scientific resources of the Nation as upon efficient industrial mobilization or skillful command of the fighting forces." [8] It should therefore be a matter of profound national concern that personnel security, when arbitrarily administered, discourages participation in military research by the very men who can supply the talent so vitally needed. The case of Dr. X, a physiologist formerly at the Edgewood Arsenal, is illustrative.

Dr. X became a member of the staff at Edgewood in 1946. At that time he had already established a reputation as an investigator of resourcefulness and high ability. For some years he had had an academic connection in which he had earned the respect of eminent colleagues. He had published some forty papers in the fields of physiology and biochemistry. During the war he participated in important studies, notably those having to do with motion sickness, under the auspices of the Committee on Medical Research, Office of Scientific Research and Development, and the Committee on Aviation Medicine of the National Research Council. Two months after Dr. X began his work at the Army Chemical Center he was curtly informed by a Military Intelligence officer that his clearance had been withdrawn; he was advised that he could resign forthwith "without prejudice" or, alternatively, he would be suspended and ultimately dismissed "with cause." At no time did he receive formal charges. A security officer, indeed, stressed that there were no charges, but that Dr. X was merely considered a "potential risk." This, he added conversationally, was because X's parents had been born abroad (though they had resided in this country at least since 1905, when X was born in New York City); because he was a member of two non-scientific organizations (neither of which has ever been cited as a "communist front" by the Attorney General or even the House Committee on Un-American Activities); because, further, he had had contact with the late Brig. Gen. Evans F. Carlson, who had aided X's wartime experiments on fatigue and motion sickness, and also with members of the faculty at a leading institute of technology, with whom he had been professionally associated; and because, finally, in 1940 he had attended a lecture given in a university hall by a gentleman who was regarded by the security officer as a "fellow-traveler."

On this flimsy foundation, without hearing or official communication of any sort other than a formal notice of suspension, Dr. X was adjudged ineligible to do the work for which he had just been recruited. Five months later, after X had submitted a self-defensive statement and an impressive array of supporting affidavits, the Secretary of the Army ordered that Dr. X be reinstated with full pay. On November 12 he was recalled to duty. On November 13 he received his salary arrears. On November 14, having been vindicated, he resigned. Since then he has been a member of the staff of a privately endowed institute.

Apart from Dr. X's personal suffering, which must have been considerable, the episode has cost the Army the services of a man who had previously been willing and apparently able to advance its researches. "Rough and ready justice" in personnel security matters is functionally unsound. The rougher it becomes the less ready are we likely to be.

Scientists Employed Privately on Military Contracts

Every contract to do research or manufacture for one of the military services contains a Secrecy Agreement. This binds the contractor to obtain written consent before he permits any alien to have access to drawings, specifications, models, and the like connected with execution of the contract. It also binds the contractor to bar the citizens in his employ from having access to any "top secret" or "secret" matters until the appropriate department gives its written consent. Employed citizens may be permitted access to "confidential" or "restricted" data without prior clearance, but this generalized consent may be withdrawn in particular cases if the military service so chooses. No distinction is made, organizationally or otherwise, between scientific personnel and any other class of nongovernmental employees.

In order to conserve manpower and to avoid conflicting decisions, the Army and Navy agreed early in 1942 that the former should execute the industrial personnel security programs on behalf of both services. When the Air Force was separated from the Army in 1947, its insistence upon an active share in administration caused a partial reconstruction of the machinery.

Under the present arrangements, an "army commander"— that is, the commanding general of an area or of the Military District of Washington—is empowered in most instances to issue a "letter of consent" if, in the light of all the evidence presented to him, he is satisfied that the employment of the individual and his having access to classified information will not be "inimical to the interests of the United States." If he is in doubt, the file is forwarded to the Personnel Security Board, which is a tripartite body composed of commissioned officers representing the Departments of the Army, Navy, and Air Force. That board decides whether consent (clearance) shall be granted or denied. Its decision ultimately reaches the contractor in the form of a letter from the appropriate army commander or, in the case of the Air Force, from the Commanding General, Air Matériel Command.

At this time, if the decision is adverse, the affected individual is notified in writing that clearance has been withheld; and he is supposedly informed, also, concerning the ways in which he may request a further review of the case by the Industrial Employment Review Board. In numerous past instances, as Army officers have candidly acknowledged, notification of appellate procedures was carelessly omitted, though there has been recent improvement in this respect. Inattentiveness to this detail was no doubt attributable in large part to the rapid demobilization and reassignment of military personnel immediately after the war, which meant that inexperienced and partially trained men were given unaccustomed tasks. The matter is of considerable importance, because neither the existence nor the procedures of the Industrial Employment Review Board have been widely publicized nor, even, made matters of record in accessible documents.[9]

The IERB is wholly separate from the Personnel Security Board. Its members have had no contact with a case before it is docketed with them for review. At that time the appropriate files are moved from the Personnel Security Board to the IERB for a fresh examination. A denial of clearance is appealable by the individual concerned (who may be represented by counsel or by his labor union if he wishes) or by the contractor-employer.

The standards of judgment for determining whether access to classified information will be "inimical to the interests of the United States" have undergone an interesting process of elaboration in recent years.

During the war years a "Joint Memorandum on Removal of Subversives from National Defense Projects of Importance to Army or Navy Procurement" defined the term "subversive activity" as meaning merely "sabotage, espionage, or any other wilful activity intended to disrupt the national defense program."

In 1946 a new effort was made to clarify the services' thinking. Administrative instructions, over the signature of General Eisenhower as Chief of Staff, dealt with "Suspension of Subversives from Privately Operated Facilities of Importance to the Security of the Nation's Army and Navy Programs." These instructions emphasized that "No employee should be suspended as a result of idle rumor, normal labor activity, gossip, or anonymous communication, *nor should an employee be suspended for any reason other than a reasonable suspicion that he is engaged in subversive activity.*" [10]

But by 1948 the emphasis that had prevailed during the war and immediately afterward was shifted. No longer was there a focus on activity as an indication of possible subversiveness. Thenceforward the test of danger was to be "a reasonable belief that the individual involved has engaged in one or more of the following activities *or associations* . . ." There then follow twelve topics, all but three of which refer to personal conduct (such as sabotage or encouragement of sedition) or characteristics (such as history of serious mental or emotional instability). The three that involve associations were stated as follows:

"(5) Affiliation with any organization or movement that seeks to alter our form of Government by unconstitutional means, or sympathetic association with any such organization, movement or members thereof;

"(6) Being influenced by or subject to the dictates of any foreign power to an extent detrimental to the interests of our Government or membership in any organization or movement so influenced by or dictated to;

"(7) Affiliation with any foreign or domestic totalitarian organization or movement or intimate or sympathetic association with any such organization, movement, or members thereof." [11]

During the twelve months from July 1, 1946, to June 30, 1947, when the stress was on "subversive *activities*," the IERB considered only three cases. During the next twelve months, toward the end of which the change from "activities" to "associations" became formally operative, the board received twenty cases. In the next two months, July 1, 1948, to September 1, 1948, thirty new cases were filed with the IERB. During 1949 approximately 110 applications for review were acted upon. Possibly the increased case load is not caused wholly by the present concern with whom a man knows rather than what he does. But a former chairman of the IERB has revealed that virtually every matter which has come before the board since April 1948 has been an "associations" case.

This shift in emphasis is a direct reflection of the "Loyalty Order," which since 1947 has been used to test the eligibility of persons who desire employment in the federal service. On November 7, 1949, the Secretaries of the Army, Navy, and Air Force formalized the relationship by issuing a new set of "Criteria Governing Actions by the Industrial Employment Review Board." In the directions they then gave the Board, the Secretaries prescribed that access to classified information should be denied for virtually the same reasons as might throw doubt upon a public employee's loyalty.[12]

The Composition of the IERB

The Industrial Employment Review Board powerfully affects the status of private persons. It determines whether they may remain in employment for which their own employers deem them to be fitted by education, experience, and aptitude.

It is of more than passing interest, therefore, to consider the structure of the tribunal.

Until the closing days of 1949 no civilian sat upon the Board whose decisions operated so directly upon civilians. All its members were military men without special training for adjudication. Four officers composed the administrative court, one voting member drawn from each military service and a nonvoting chairman who was detailed to that duty by the Provost Marshal General of the Army. The chairman, despite his inability to vote, was from the first the true director of the Board's operations. He organized the evidence, conducted the major portion of the questioning during hearings, and formulated the decisions that were reached. No member of the Board, not even the chairman, devoted full time to its work; each of the members except the chairman had as an alternate a brother officer who could sit in his stead when he was otherwise occupied.

The military cast of the Industrial Employment Review Board was strongly criticized during 1949 by the American Association for the Advancement of Science, the American Civil Liberties Union, and others. Partially in response to these promptings a significant change was initiated in December 1949. The IERB was removed from the Office of the Provost Marshal General of the Army. It has been reconstituted as a joint board of the Departments of the Army, Navy, and Air Force, responsible to the Secretaries of those departments rather than to a general officer. Its policies are to be framed or approved by the Munitions Board, a civilian agency within the Department of Defense, which, moreover, has been empowered to appoint a civilian as the IERB's chairman. The members of the IERB (and their alternates) other than the chairman are to be appointed by the respective military Secretaries and may be either officers or civilians. At least one member of the Board must now be a member of the bar. Three members of the Board constitute a quorum, but the lawyer-member must be one of the three when a case is being finally decided.

The reorganization of the IERB in terms that give it a somewhat civilian rather than an exclusively military orientation is no reflection upon the officers who have previously staffed the tribunal. They have served conscientiously and, especially in the case of the successive chairmen appointed by the Provost Marshal General, have been reasonably aware that civil rights as well as military security are important to the nation. Occasionally there have been intimations of occupational attitudes that are perhaps irrelevant to the task at hand, as when the Army member in a belligerent and hectoring tone of voice demanded to know why a young scientist had not been in uniform during the war. When the young man mildly replied that his employer had sought his draft deferment because of the importance of the work he was then doing, the officer sneeringly snapped the question, "Nobody stopped you from enlisting, did they?" This sort of occurrence, however, has seemingly been rare; there has been little reason to challenge the Board's members for having blustered or having been willfully blind to favorable evidence. In more cases than not the Board has reversed the unfavorable action of the security officers and has directed that clearance be granted; according to one member of the IERB, it has learned that officers who spend a great deal of time in investigating charges of subversive associations "tend to develop fixations and only look at the bad side of the record." [18]

Nevertheless, there is considerable ground for arguing against the further appointment of officers to sit on this tribunal. The IERB, as a body that determines the economic and social fate of civilians by adjudicating their professional or occupational opportunities, ought to be composed entirely of civilians, answerable to other civilians rather than in part to military superiors. The body is essentially a judicial one. Service upon it is not a very rewarding side-line activity for a professional soldier. No military knowledge is involved in its deliberations. Whether the material to which access is sought should be classified as "top secret" or "secret" or "confidential" is not a question before the Board. Appropriate officers of the National Military Establishment will already have considered that problem, and will have settled it authoritatively. The IERB concerns itself solely with the citizen's reputation and reliability. An issue of that sort is not within the specific and distinctive competence of the military. As a matter of important principle, members of the Industrial Employment Review Board should be selected from those who are not in the active service of the Army, the Navy, and the Air Force.

This principle is buttressed by one of our most deeply rooted national traditions. From the earliest days of the republic's existence, the American military has been subordinated to civil authority in other than strictly military affairs. Indeed, one of the grievances listed in the Declaration of Independence was that the British had exalted the military over civil power in the American colonies; the first constitutions of Delaware, Maryland, Massachusetts, North Carolina, Pennsylvania, and South Carolina specifically reversed the allocation of control by providing that civil authority should at all times prevail over the military. General Washington himself declined to try civilians before military tribunals until he received express authorizations from the Revolutionary Congress. And in later years the Supreme Court has held with great consistency that action by military authorities having impact upon private rights cannot be sustained merely on the ground that there is "military necessity" for them; only a danger that is "immediate and impending and not remote and contingent," or a specific Congressional authorization, can serve to blur "the boundaries between military and civilian power," which, the Supreme Court recently said, have "become part of our political philosophy and institutions." [14]

This is not merely a matter of interest to legal antiquaries. The history of all the world shows that truly dominant militarism has grown out of the gradual and often even unintentional absorption by the army of state functions, and the performance by soldiers of duties for which military service provides no peculiarly useful equipment. Tribunals manned by officers whose profession is arms rather than justice have in the past produced bizarre conclusions arrived at after shocking procedures.[15] It is praiseworthy that superintendence over the functions of the IERB has, by the recent reorganization, been committed to a civilian chairman before the true nature of those functions had become entirely obscured by reason of their having been too long performed by soldiers. The Secretaries of the military departments now have it within their power to supplant the remaining military members. Thus far they have shown no disposition to do so. All but the chairman of the Board are officers, while other officers serve as the tribunal's executive director and procedural adviser. Competent and disinterested as they no doubt are, they should now be replaced. Happily, no real or supposed threats to public order require abandoning fundamental procedures or reshuffling the division of power between civilian and military authorities in the United States.[16]

Centralization of IERB Proceedings

The Industrial Employment Review Board is not, from the standpoint of the individual involved in its proceedings, a true review or appellate body. It is, rather, the trial court. At no earlier point has the affected person had opportunity for interview or for hearing, whether formal or informal. At no earlier point, in fact, has he had even a generalized notice that his status is in question. When, therefore, he is advised that clearance has been denied by the commanding general (who, as we have seen, acts upon the advice of the Army-Navy-Air Force Personnel Security Board), he turns to the IERB for the trial hearing he has not yet had.

In one important respect, however, the IERB is unlike other courts of first instance. When a trial court speaks, it renders a judgment which, ordinarily, is subject to review by some higher tribunal. Not so the Industrial Employment Review Board. It is not only the court of first resort. It is also the court of last resort. Its verdicts are final and unreviewable.

Few systems of law administration in modern society have failed to provide opportunity for correction of errors by the tribunal that first hears a case. The absence of such an opportunity in this instance is made more serious by the fact that the IERB sits exclusively in Washington. This centralization results, as a practical matter, in denial of hearings in many cases. The expense of attendance upon sessions in Washington effectively prevents appellants from presenting their defenses in person or through counsel of their own choice. Witnesses cannot be transported except at a cost that makes it unfeasible to offer their testimony. True, the chairman of the Board or one of its members occasionally leaves Washington in order to hear matters which have arisen at distant points. In no case within five hundred miles of Washington, however, has there been a chance to obtain a cheap and convenient hearing. And even when a Board member does "ride circuit," the appellant may still have long distances to travel before he reaches the assigned place of hearing. Once arrived, he faces only a single officer rather than the Board as a whole; the final decision is handed down by the Board in Washington on the basis of the stenographic transcript and the presiding officer's oral recommendation, which is undisclosed to the appellant and which he can do nothing to counter.

A start has been made in overcoming these very real organizational difficulties. The new charter of the IERB permits the creation of regional or area boards, composed as is the central body and possessing the same measure of authority. As yet no regional boards have been created, and responsible officials have privately stated that none will be unless the present case load should unexpectedly increase. Plans are under way, however, to designate referees or trial examiners who will be able to conduct proceedings locally, not with a view to making decisions, but merely to permit a hearing to be held in a suitable place without intolerable expense to the appellant. It is not now contemplated that the referee will do more than take testimony. The resulting record will be forwarded to Washington for authoritative consideration by the IERB itself.

If these plans mature, they will make for improvement in the present situation. But they do not go quite far enough.

The first step must be to recognize the IERB for what it is, namely, a trial board rather than an appellate board, and then to replace it with a true review board to which unfavorable judgments may be appealed.

The second step must be to provide trial boards that will sit in or near the major industrial and educational centers of the country as occasion may arise. Only in that way can the opportunity for hearing become a practical reality in all cases, rather than a mere form of expression. Like the reconstituted review board, the trial boards should be composed of civilians; distinguished citizens could very probably be readily recruited for this part-time public service. A working model is at hand in the personnel security operations of the Atomic Energy Commission. There, it will be recalled, local boards hear the cases in the first instance, subject to later review by a central appellate body. The AEC model may be suggestive of one additional improvement in military security matters.

In AEC cases involving a professional employee, a member of his profession sits on the trial board. This affords desirable assurance that the hearings will not ignore the bearing of the respondent's work on the whole project of which he is a part. A similar occupational representation in military security cases might reinforce sobriety of judgment and might encourage general confidence in the fairness of the proceedings.

V

The Spreading of Security

Requirements

THE world being what it is, one would be naïve indeed to assume that American laboratories are immune from espionage. On June 20, 1949, the President of the United States signed the Central Intelligence Act of 1949. Its purposes and implications were deemed to be so confidential that the House and Senate Armed Services Committees discussed the measure in secret sessions. Then they informed Congress that it was impossible to have a full debate or even to disclose the objects and operation of the proposed statute. The Act provides in part that the Director of Central Intelligence shall have the power, upon his own certificate and without regard to any other laws relating to public expenditures, to spend sums for "objects of a confidential, extraordinary, or emergency nature," without any review of his acts by the Comptroller General, the Bureau of the Budget, or Congress itself. The Act also empowers the Director to appoint highly qualified personnel in order to effect "scientific intelligence functions relating to national security." There is really no reason to suppose that other countries are any less interested than are we in "objects of a confidential, extraordinary, or emergency nature" and in "scientific intelligence functions." Whatever may be the euphemisms in current use, it is obvious that today as in the past the major powers seek to spy on one another, whether the spying be done in Canada or the United States, or, perchance, the Soviet Union. And so long as we fear that spies may masquerade as scientists, it is understandable that resort will be had to screening processes like those outlined in the immediately preceding chapter.

If the screening were confined to those who had access to classified materials or to restricted areas, the matter might be checked off as simply another of the unpleasant costs of war, cold or hot. In fact, however, official inquiries into individuals' "reliability" goes so far beyond these limits that an entirely new policy question is raised. That question, in sum, is whether the nation gains by extending security clearance requirements, or their equivalent, to large numbers of scientists who are not themselves engaged in classified research projects and who neither need nor have opportunity to acquire secret information.

A few examples will illustrate the dimensions of this new problem.

Not long ago, an eminent British scientist was refused permission to visit American universities having *unclassified* military contracts in his area of specialization, unless he first underwent the conventional clearance procedures. This he could not do within the time limits of his stay in this country. According to a recent report to the State Department, "This man is conservatively estimated to be fully two years ahead of his American colleagues with respect to his field. Hence his visit was largely an opportunity in which American science had everything to gain with little to return. The further researches of our own people, deprived of the opportunity of making a two-year forward step in their work, represent the subsidization of an inferior effort." [1]

The Brookhaven National Laboratory is administered by

nine eastern universities, Columbia, Cornell, Harvard, Johns Hopkins, Massachusetts Institute of Technology, Pennsylvania, Princeton, Rochester, and Yale, which have banded together for this purpose into Associated Universities, Incorporated. While all of Brookhaven is devoted to Atomic Energy Commission projects, most of the work done there is entirely unrestricted. Brookhaven's radiological and nuclear research is mainly of a basic nature, though it is by no means "purely theoretical." The projects involve such varied matters as designing new particle accelerators, studying the effects of irradiation upon the functioning of the endocrine glands, measuring radioactivity in the atmosphere, exploring the effects of gamma radiation upon various field crops, and investigating the ways in which the human body utilizes iron and other metals. Enterprises of these sorts account for perhaps 90 per cent of Brookhaven's activity and staff. Only the remaining 10 per cent of the work is classified, because it involves knowledge of the planning, properties, and performance of the uranium-graphite reactor that has been built at Brookhaven as a research tool rather than as a large-scale producer of fissionable material.

The Atomic Energy Act of 1946 mandatorily prescribes the clearance of the few Brookhaven researchers who may have access to restricted data. Nothing in the law, however, requires that all the other scientists in that large installation be cleared as a condition of their being employed. Yet it appears to be true that in the past local officials of the AEC have, in the words of one of Brookhaven's administrators, "strongly intimated" that all scientific personnel should be cleared, regardless of the nature of their work. During one brief period of time five out of eight men who had been proposed for appointment to posts involving pure theory and no restricted data were denied clearance and were not employed even though other qualified men could not be recruited in their stead. More recently this semiofficial encouragement of indiscriminate enforcement of clearance requirements has abated, apparently not so much as a matter of conviction as because it imposed too heavy burdens upon the Federal Bureau of Investigation and the AEC staff, which were compelled to "process" the many cases involving no access to restricted data and hence beyond the scope of their statutory duties.

Now, however, the Associated Universities, Incorporated, voluntarily continue at least a part of the practice that had been unwisely inspired in the beginning. Without obvious prodding by AEC officers (who, incidentally, seem not to have been reflecting any policy formally established by the Commission itself), the Brookhaven administrators still seek clearance for all scientists who are to be stationed at the laboratory more or less permanently rather than merely for temporary duty as, for example, are many university professors and graduate students. The declared reason for the present position is that these scientists may at some future time desire to use the reactor or to consult classified materials. The reason lacks persuasiveness because one side of the reactor is to be declassified, thus making it available to qualified researchers without clearance.

Brookhaven is not alone in pursuing this policy. Industrial laboratories, too, have demanded clearance as a prerequisite of employment, even though the classified work in those laboratories may require the services of only a small number of the scientists who are employed at any one time. This is the case, for instance, at the General Electric Company, although the classified researches that it has undertaken to do under contracts with the Government are physically separated from the rest of its scientific operations, and despite the fact that GE's basic research laboratory had traditionally had an open-door policy. Here again the justification advanced is one of "administrative flexibility." Even though a man is not engaged in classified research, it might be desirable at some future time to transfer him to it. If everyone is cleared in advance, the reassignment of staff is made easy. Similarly, in Oak Ridge the Carbide and Carbon Chemicals Corporation makes no differentiation between those of its employees who have access to restricted data and those who do not. Michael F. McDermott, the company's Superintendent of Security and Plant Protection, writes: "Carbide, in requesting clearances from AEC, has in all cases sought 'Q' clearances on the theory that while one's designated job may be in a limited area [in which unclassified work is done], there may be times (through visits, transfers or possibly visits to other projects outside of Oak Ridge) when one would be subject to classified data and would then have been cleared for such visits or information."

A third situation, which will be considered more fully in a subsequent chapter, is perhaps yet more alarming. Academic institutions in increasing numbers have manifested an interest in security clearance before making an appointment to teaching or research staffs. This self-created limitation upon institutional freedom has seemingly been induced chiefly by a desire to obtain research funds from federal agencies.

All these extensions of clearance beyond the true scope of security administration are influenced in some measure by a suspicion that lies just below the surface of public consciousness. J. Robert Oppenheimer some years ago asserted in an off-hand and unfortunately quotable way that the best method of transmitting scientific information was to wrap it up in a person. One would exaggerate the significance of the comment if he ascribed all later developments to it. Nevertheless it seems clear that the broadening of clearance is inspired by something more than considerations of mere administrative convenience. It is inspired as well by the fear that one scientist may talk to another outside the laboratory. If he does so, he may communicate information that will become wrapped up in a person to whom it should not have been disclosed and who may in turn transmit it to unauthorized recipients. "We all believe," once remarked the manager of the AEC's New York Operations Office, "that it is unwise to have unreliable men working with those who are doing classified work. There is no label on a man to indicate the nature of his work, and scientists are a tight community." Somewhat similarly, the General Electric Company's director of research activities has upheld security clearance for the whole staff of a laboratory, classified and unclassified alike, on the ground that it permits all the scientists to talk freely with one another without having to fear "leaks."

The trouble with this sort of suggestion is that it proves at once too much and too little. Perfect security cannot be achieved by extending clearance merely to the "tight community" of scientists in any particular laboratory. The researcher at the various AEC installations is often an academic man who has accepted a short-term assignment. He corresponds with and will soon rejoin his faculty associates. By logical reasoning must one not conclude that they, too, should be cleared lest some of them prove to be "unreliable" and eager to corrupt their colleague, the possessor of classified information? Equally, would it not be necessary to clear everyone with whom an industrial scientist had repetitive relationships, lest his conversational excursions over cocktail glasses contain inappropriate references to his work? A scientist who is designedly or carelessly unmindful of his obligation to avoid revealing restricted data may unauthorizedly transmit information to anyone he knows. If we permit ourselves to be consumed by dread of that possibility, we must either extend security clearance to all who may meet a scientist or, alternatively, must prevent our "cleared" scientists from having contact with the "uncleared" world which surrounds them.

No one is attracted by such drastic extremes. Their unpalatability is heightened by realization that they are not necessary in fact and, even in their present incomplete approximation, are hurtful to the very cause they are intended to serve, national security.

Of course the fact that scientific espionage has not been dramatically successful in the past does not mean that it will certainly be a failure in the future. It would be foolhardy to take no precautions whatsoever against improper communication of restricted data. Prudent men take precautions against even slight risks. But unless the risks are grave, prudent men do not live constantly in the shadow of fear.

What does the public record show which sheds light on the gravity of the risk that scientists as a group are not quite reliable? It shows, to be sure, that scientists have been among the espionage agents of foreign powers. But the number of spies, so far as we know, has indeed been small—May and Boyer, an Englishman and a Canadian, in Canada; Fuchs, a British citizen, who transmitted intelligence both from England and from this country in which he was temporarily stationed during World War II; Gold, an American chemist not employed in government work at all, who was the conduit used by Fuchs; and a limited group of relatively minor figures who were apparently also on Gold's team during the war years.

Corruption and faithlessness, no matter how infrequently they occur, can never add up to a pretty story. It would be a mistake, however, to attach to a large and devoted profession the repugnance engendered by a few isolated cases involving individuals within that profession. It is impressive that not a single one of the scientists involved in security clearance proceedings during the years of Russo-American tension since World War II has been found to be a spy, either amateur or professional. No basis appears for manifesting an especially distrustful attitude toward American men of science.

The House Committee on Un-American Activities has sought, perhaps successfully, to create contrary impressions. Analysis of its reports demonstrates, however, that their substance is slight indeed. Consider, for example, the Committee's utterances that evoked the following sequence of page one headlines in the calm *New York Times* during September 1948:

September 2: "HOUSE BODY TO SIFT SPYING FOR RUSSIA BY ATOM SCIENTISTS"

September 8: "WITNESS CALLED FOR ATOMIC INQUIRY
Secret, Open Hearings Are Set on Scientific Project 'Leaks' "

September 18: "HOUSE BODY PLANS TO EXPOSE DETAILS OF ATOMIC SPYING"

September 25: "PUBLIC SPY INQUIRY OFF; 'GRAVEST MATTER' UNCOVERED"

September 26: "ATOMIC SPY REPORT WILL SHOCK PUBLIC, OFFICIAL DECLARES"

September 28: "INDICTMENT OF FIVE URGED IN REPORT ON ATOMIC SPYING
House Group Lists Two Scientists as in Bomb Project"

When the Committee's report was released, it became clear that neither of the scientists specifically denounced by the Committee had been connected with the "Bomb Project" for a number of years. They had been employees in the days of the Army, rather than the AEC, and early in the game they had been called to active military duty far from any laboratory because the Army doubted their probity. Even as to these two the Department of Justice on September 29, 1948, issued an official commentary on the House Committee's report, asserting that the Department had no evidence that warranted prosecution and stating in part:

"... There is absolutely no competent proof here, so far as appears from the report and excerpts of testimony quoted therein, of the actual or attempted communication, delivery or transmittal of information relating to the national defense to a foreign government or to one of its representatives ... The Committee ... has uncovered nothing the department did not already have. ... It has been the considered judgment of two successive assistant attorneys general who studied the facts available, independently and at different times, that the evidence was insufficient for successful prosecution ... The Congressional 'reports' on espionage and loyalty matters ... are injurious to the principles of free government. ..."

Nevertheless, the succession of sensational Committee news releases undoubtedly aroused in many unsophisticated minds a feeling that scientists who work on the atomic energy project are a pretty doubtful lot.

This feeling has been fully exploited by repeated announcements concerning "atomic spying by scientists," each announcement sounding like a fresh revelation. All of them have, however, involved the mere repetition of allegations against the same three individuals, two of whom were those involved in the headline series quoted above, and the third of whom was known until recently simply as "Scientist X." The accusations against "Scientist X" were released by the Committee on at least three widely separated occasions in precisely the same words. On the third occasion the Committee's pronouncement was still treated as "fresh news"; on August 16, 1949, the *New York Times* carried a headline, 'SCIENTIST X' LINKED TO ATOMIC ESPIONAGE, eleven months after the Committee had first given out the identical story on

September 28, 1948. Finally, on September 29, 1949, the Committee issued a "Report on Atomic Espionage" that told the story once again, the new item being that Scientist X was now identified by name and occupation as a midwestern university professor. All the Committee's allegations related to 1943 or earlier years, and in no instance has the Committee's evidence yet been deemed sufficient by the Department of Justice to support a criminal prosecution.

As for scientists' membership in the Communist Party, with its implication of conflicting loyalties, the Committee has reported what it described as a cell "consisting of five or six young physicists" who were connected during the war with one or another phase of work at the Radiation Laboratory at the University of California, a part of the atomic energy project. None of those named by the Committee has been found guilty of any misconduct in connection with the project. Only one of the group had continued in his employment beyond the war, and well before his public "exposure" by the House Committee the AEC had demanded and received his resignation. According to testimony before the House Committee, he had been a Communist for three months, from January until March, 1943, had paid fifty cents in dues, and had then withdrawn from the party; during this period he had been "a computor, a mathematical computor. I worked a little electric gadget, pushing buttons." [2]

The House Committee's penchant for repetitive denunciation has apparently befuddled unwary readers into supposing that there are more cases and more proofs than in fact exist; evidence is at hand, moreover, that newspapers with dominatingly large circulations have tended to be especially generous in reporting exclamations by the Committee, its members, or its staff.[3] It is not too much to say that the loyalty of scientists as a group has become a matter about which there is wide public concern.

For this reason it is well to emphasize once again that security clearance cases have in truth rarely involved charges that a scientist is himself a "disloyal" person. Without now pausing to describe the charges in detail, we may characterize them generally as involving dissatisfaction with an individual's associations. Sometimes the associations have been entirely personal, as, for example, a case in which a scientist's father-in-law had once been the editor of a Yiddish newspaper of an allegedly radical character; the theory seems to have been that he might have "infected" his daughter, who might in turn therefore be an unwholesome associate for her husband, the scientist against whom the inquiry was directed. Sometimes the associations have been of a professional nature, as, for example, a case in which a distinguished consultant was challenged because two of his colleagues on a university faculty were asserted to be "Communist sympathizers." Sometimes, finally, the associations have been with organizations in which Communists are said to have been active, as, for example, a case in which a young scientist's clearance was long delayed because he had once joined a "United People's Action Committee" for the declared purpose of combating discrimination against Negroes in Philadelphia.

In all instances of these sorts, the root proposition is that the scientist might be indiscreet in the presence of his associates or that he might at some time be induced by them to perpetrate an illegal act, such as divulging secret information. This proposition is not wholly irrational. The chief criticism to be made of it is, simply, that it impliedly assumes a degree of danger that does not exist. Dr. Leland T. Haworth, the director of Brookhaven National Laboratory, recently formulated this thought in a mathematician's terms:

"I suppose that there is always a risk that a man may break security, because he is disloyal or otherwise. The probability of any particular scientist's being loyal is not infinitely great. So let us suppose that there is one chance out of ten that a man will be disloyal—certainly a higher supposed probability than the facts seem to warrant. Then let us suppose that the chances of there being any useful information at Brookhaven are about fifty-fifty and that the chances of any particular person's being able to lay his hands on the desired data are three out of ten. The chances of his being able to get the information to some other country in any useful form are surely not more than one out of a hundred. Multiply all these together—$.1 \times .5 \times .3 \times .01$—and you get .00015, or less than two chances out of 10,000 that information arrives where we don't want it to go. Of course there is always a possibility that a man's mother-in-law or Great Aunt Sally will pick up some useful data from him, but the possibility is so exceedingly small that we ought to disregard it. When we bring up these flimsy 'associations' charges, we're likely to lose more than we gain. We lose the talents of the suspects and we scare hell out of the rest."

This forcefully stated conclusion finds ready support in the known facts. The right to be let alone by the Government, as Mr. Justice Brandeis put it, is "the most comprehensive of rights and the right most valued by civilized man." [4] A number of capable and personally irreproachable scientists who value that right have simply withdrawn from important research positions because they reasonably feared that their relatives or friends, not they, would be smeared in clearance proceedings. Others for similar reasons have declined invitations to undertake assignments of national importance; in instance after instance those who are responsible for recruiting men for the more advanced jobs have confirmed this observation, some of the estimates rising as high as 50 per cent, though most have been substantially lower. Still other well equipped persons, having been recruited for work of a non-secret though nevertheless significant nature, have become discouraged by delays in obtaining clearance and have accepted other employment instead. At Oak Ridge and elsewhere, moreover, academic men who have been willing to do work during the summer months have often been prevented from engaging in it because their clearance was not granted until the summer was almost at an end. The numerical aggregate of these losses is considerable.[5] Their significance is even greater than their number because there is not an inexhaustible supply of trained scientists who can be found overnight to replace them.

This fact acquires especial importance because of an ever-growing tendency to avoid recruitment of men and women who might conceivably encounter "clearance difficulties." Many scientists, though already cleared themselves, hesitate to recommend the appointment of a fellow-scientist whose general outlook is thought to be "liberal." Their reluctance to do so is in part the product of tender concern for their scientific friends, whose reputations might be damaged if they were not cleared. In part it is the product of fear for themselves; a man's own reputation may be damaged if his friends have clearance difficulties, for this will immediately suggest that the nominator himself has questionable associations. Needless to say, this dual timidity produces many errors on the side of caution, and thus immeasurably broadens the range of ineligibility.

Two illustrations suffice to make this point. Not long ago there was undertaken an important survey of the medical research facilities of the military departments. Those who planned the project drew up a roster of outstanding medical scientists to conduct the survey. Then it was observed that one of the nominees, a man especially fitted to give advice in one of the key phases of the survey, was reported to have been a supporter of Henry Wallace. Without his even knowing that he had been considered and rejected, he was promptly stricken from the list and a second choice made. At no time was there any denial of clearance; the name was simply never submitted, in order to forestall the feared embarrassment of clearance proceedings. In all probability there would have been no withholding of clearance if the case had been presented and if the facts had been as supposed; so far as I know, there has not yet been an adverse decision because in 1948 a scientist voted for Mr. Wallace instead of for Mr. Truman or Mr. Dewey or Mr. Norman Thomas. But the rather sour cream of this jest is that the man in question was not a Wallace supporter at all; he had been a sturdy upholder of the Democratic candidate. In sum, the military department was deprived of the services of an eminently qualified and badly needed adviser because of a conscientious but erroneous assumption concerning a fact that was irrelevant in any event. In a second case the occupant of a responsible scientific post in the federal service had recently received his own security clearance after it had been brought in question for rather unsubstantial reasons. Just at that point a vacancy arose in his staff, for which there were two applicants. One, who had had the more extensive experience and who possessed an already established reputation, had freely voiced opinions which, though wholly American, are not acceptable in every quarter. In these circumstances the second applicant was chosen for appointment. Of course the same choice might have been made even if the circumstances had been different, but the decision was very probably a recoil from the proceedings through which the appointing officer had himself so recently gone.

Unless the fear of smear can be pushed farther into the background than it is at present, the skills of many of the nation's ablest scientists will not be fully utilized. Ebullience and unorthodoxy may not be absolute prerequisites of scientific brilliance, but they are certainly compatible with it and probably accompany it more often than not. Authoritarianism is by definition inconsistent with intellectual creativeness. The questing scientific mind does not embrace theories without proof, and even accepted theories remain always subject to possible broadening or modification. "The history of ideas," says Alfred North Whitehead, "is a history of mistakes." The entire history of science has been one of battle against orthodoxy—against the orthodoxy of the church, the orthodoxy of dogmatic conviction and intuitive knowledge, the orthodoxy of social, economic, and political opinions of the moment. Although the habit of doubting the perfection of things as they are may sometimes be indulged unwisely, it is almost a necessary attribute of those who contribute to the progress of science. The scientific drudge may live untouched by the turmoil of ideas. The scientific creator is likely to be broadly cultured, complex, alert, and unafraid of the unconventional. Too many men who possess those characteristics are today avoiding work for which clearance must be sought, or are being passed over in favor of more pedestrian spirits.

Sumner Pike, one of the Atomic Energy Commissioners, has cautioned against just this possibility. "The degree of success in our job," he said, "depends fundamentally on considerable numbers of scientific minds of the highest quality to carry on exploration into unknown or dimly perceived fields of research. Such minds must be brilliant, curious, skeptical, and roving. They do not take things for granted. They must examine and re-examine conclusions reached by others before reaching their own decisions." We need many more of that type, adds Commissioner Pike, even though their being accustomed to scientific freedom of expression tends to make them "outspoken on social injustices and unnecessarily tactless in exposing our own troubles here at home." [6] These opinions have been interestingly and perhaps surprisingly echoed by a veteran member of an AEC personnel security board, who attributes his initial selection to "a New England Republican background and the fact that I had never bothered to think about social and political questions." In the course of his service as security-risk judger, he has concluded that "The men whose clearance status is called into question are usually those who do think about these things, the very kind of men you want on this job. In this business many of the ideas start at the bottom rather than the top, and ideas are likely to grow out of active minds rather than those which accept things just as they find them."

The ends of true national security, as these remarks once again emphasize, are not served by confusing orthodoxy with suitability for scientific service.

Personnel clearance, as earlier pages have indicated, is defensibly prudent when confidential assignments are involved. When a man's acts may heavily affect the community's safety, a judgment concerning his probable future conduct may appropriately be made, even though the judgment is perforce inexact. In such a case society balances risks. On the one hand there is a risk that infidelity may cause grievous injury to the nation. On the other hand there is a risk that an erroneous conclusion about an individual may be grievously injurious to him. It is not unreasonable to conclude that the first of these risks outweighs the second, and that personnel security determinations are therefore well justified. The justification, however, is related to and derives from the existence of potentially grave danger. If danger is in fact not present, or if its degree is inconsiderable, the stated justification vanishes. The extension of personnel security clearances into areas in which they are not demonstrably necessary protects no national interest.

VI

The Loyalty of Federal Scientists

THE previous discussion has dealt with the tens of thousands of scientists who are employed in activities in which secrecy and security are thought to be important issues. As has been seen, great effort is made to guard against employment of unsuitable personnel in work of that kind.

Wholly unrelated to the "sensitive areas" that have thus far been considered, some thirty thousand civilians have professional civil-service ratings in federal agencies as chemists, physicists, meteorologists, entomologists, geologists, bacteriologists, pathologists, astronomers, and so on. To that number must be added the many thousands of supporting technical personnel and the yet further thousands of doctors, dentists, psychologists, and the like who are employed by the Veterans Administration, the Public Health Service, and other departments. Even those scientists who do have access to restricted data possess, for the most part, few real secrets—certainly far fewer than many normally self-assertive men ever permit their acquaintances to suppose. As for the scientists who will be discussed in the present chapter, there is no room whatsoever for speculation on this score. They are factually, officially, and unqualifiedly barren of state secrets. They have not the slightest opportunity to deal in restricted data or to magnify their own importance by multiplying the number of hushes in hush-hush.

The inconspicuous ichthyologist of the Fish and Wildlife Service knows many secrets, to be sure, but they are the secrets of the speckled trout rather than the secrets of national defense. The mine safety engineer in the Department of the Interior peers into dark and hidden places, but the information he acquires has no element of confidentiality. The researcher at the National Cancer Institute explores the unknown, but there is certainly no disposition to conceal whatever he may discover. The Liberian scientific mission of the Public Health Service and the Agriculture Department is engaged in work of national importance, but whatever it learns about *Strophanthus sarmentosus* as a ready source of adrenocortico trophic hormone will not be withheld from the rheumatoid arthritis sufferers of the world. When Dr. Elmer W. Brandes of the Bureau of Plant Industry proved that "yellow stripe," which once threatened the sugar cane industry with extinction, was a virus carried by the corn louse, his work was recognized to be of international significance; but no one was disturbed by the knowledge that Dr. Brandes would not "keep the secret." Dr. Ralph R. Parker of the Rocky Mountain Laboratory of the National Institutes of Health devoted long study to the wood tick and to the spotted fever which it spread with often fatal consequences; the effective vaccine that resulted from his researches was a cause for rejoicing and acclaim rather than for silent concealment. Dr. Charles A. Cary of the Bureau of Dairy Industry discovered a nutrient in milk, later identified as Vitamin B12, which helps overcome pernicious anemia; there was no fear that security would be jeopardized if he were to publish his findings, even though they might be translated into Russian. No more was there a feeling that secrecy should be clamped upon Edgar S. McFadden's recent development of a rust-resistant wheat.

Yet the political views and the associations of all these men, and of others like them, have been a matter of governmental scrutiny almost as though they were entrusted with the latest developments in chemical warfare or rocket design.

The Loyalty Order

On March 21, 1947, President Truman proclaimed that "the presence within the Government service of any disloyal or subversive person constitutes a threat to our democratic processes" and that "maximum protection must be afforded the United States against infiltration of disloyal persons into the ranks of its employees." Accordingly the President on that day promulgated an order—Executive Order No. 9835— "prescribing procedures for the administration of an employees' loyalty program in the Executive Branch of the Government."

By the terms of that decree, every person in the employ of, or seeking to be employed by, any department or administrative agency of the Federal Government must be subjected to a thorough "loyalty investigation." The Loyalty Order, as Executive Order No. 9835 has come to be known, establishes that "The standard for the refusal of employment or the removal from employment in an executive department or agency on grounds relating to loyalty shall be that, on all the evidence, reasonable grounds exist for belief that the person involved is disloyal to the Government of the United States." This is entirely in addition to, rather than a substitute for, the statutes and regulations which, in aid of national defense, authorize the summary dismissal of employees of the State Department, the Atomic Energy Commission, the Central Intelligence Agency, and the several military departments.

Each employee and each new job applicant must file his fingerprints and must answer under oath a detailed personnel security questionnaire, or PSQ. A summary check is then made of available records—Civil Service Commission, Federal Bureau of Investigation, Military and Naval Intelligence, House Committee on Un-American Activities, and similar pertinent sources—to discover whether any "derogatory information" appears in connection with the applicant's or employee's name. If none is discovered in these files or in the completed PSQ, the FBI, which is the agency in charge of conducting all loyalty investigations, reports that no derogatory information has been found; and there the matter rests. If even a minute amount of derogatory information does appear, a "full field investigation" is undertaken by the FBI, which ultimately turns over its report to the employing agency or, in the case of new employees (who are defined as all those employed after October 1, 1947), to the Civil Service Commission.

In each department or agency one or more loyalty boards, composed of at least three departmental officials, have been designated by the agency head to pass on loyalty cases affecting present employees, while in each region of the Civil Service Commission there has been created a Regional Loyalty Board to consider the cases of new employees. If an agency's loyalty board makes an adverse recommendation, the affected employee may appeal to the head of the agency or to his designee, and the agency's final decision (or, in the case of a new employee, the decision of the Regional Loyalty Board) is appealable to the Loyalty Review Board of the Civil Service Commission. Formally the decisions of the Loyalty Review Board are advisory only, but the President himself has indicated that in fact they are to be deemed virtually dispositive. The Loyalty Review Board, composed of prominent citizens under the chairmanship of former Assistant Attorney General Seth W. Richardson, has issued numerous "directives" to the several agency boards, and in general comports itself as though it were the immediate administrative supervisor of the other units in the loyalty program.

Guides to Disloyalty

The issue before these administrative tribunals is whether "reasonable grounds exist for belief that the person involved is disloyal to the Government of the United States." What are the criteria by which so elusive a matter is to be judged?

The Loyalty Order itself describes various "activities and associations of an applicant which may be considered in connection with the determination of disloyalty." These include:

1. Actual or attempted sabotage, espionage, treason, or sedition;
2. Advocacy of revolution or force to change the constitutional form of government of the United States;
3. Intentional and unauthorized disclosure of confidential documents or information obtained as a result of public employment; and
4. Performance of duty "so as to serve the interests of another government in preference to the interests of the United States."

Obviously enough, these offenses can be established only by objective evidence of actions already taken and of deeds committed. Without exception they refer to behavior rather than to belief or emotion. A significantly large volume of penal statutes applies to these types of acts, as well as to conspiracies, combinations, and attempts to commit them.[1] It is clear, moreover, that in cases where the available proofs might not be sufficiently clear-cut to sustain a criminal prosecution, the employee would nevertheless be subject to removal from his job. Like other employers, the Federal Government has a comprehensive power to dismiss or otherwise discipline an employee

131

who is insubordinate, incompetent, or inattentive to the policies he has been instructed to administer. Even employees who, in the inexact but popular phrase, are "protected by civil service" are still subject to being removed "for such cause as will promote the efficiency of the service"; and in the remaining employments to which civil-service laws are inapplicable, there are no significant limits whatsoever upon administrative discretion. Long before the Loyalty Order was born, there was ample power to take protective steps against an employee who was believed to be a saboteur, a sieve through which confidential information passed, or a servant of another nation's interest. The Order somewhat elaborated the procedural steps that were to be taken if this kind of case arose, but it added nothing to the content of the safeguards against these types of misconduct. It is proper to conclude, therefore, that the Loyalty Order was not devised to cope with behavior of these sorts.

Nor was the Loyalty Order at all needed in order to authorize the Government to rid itself of Communists. The Hatch Act, which became law in 1939, provides that no person may be employed by the Federal Government in any capacity if he has "membership in any political party or organization which advocates the overthrow of our constitutional form of government"; and as though to underline its resolve, Congress has inserted in all general appropriation acts since 1941 a reminder that no part of the appropriation may be used to pay the salary of any "person who advocates, or who is a member of an organization that advocates the overthrow of the Government of the United States by force or violence." [2] The Attorney General has unequivocally ruled that by virtue of these laws members of the Communist Party, the Socialist Workers Party, and the Workers Party are instantly removable from any post in which they may be found.

What the Loyalty Order has freshly supplied as a possible

reason for dismissal may be found in a single subparagraph, the final one of the "standards" listed in the Order and the one which generates most of the loyalty charges:

"f. Membership in, affiliation with or sympathetic association with any foreign or domestic organization, association, movement, group or combination of persons designated by the Attorney General as totalitarian, fascist, communist, or subversive, or as having adopted a policy of advocating or approving the commission of acts of force or violence to deny other persons their rights under the Constitution of the United States, or as seeking to alter the form of government of the United States by unconstitutional means."

Membership is a clear concept, and affiliation, though less precisely ascertainable, has also been given meaning by Supreme Court definition. The acts that tend to prove "affiliation" with a group may be intermittent or repeated, but according to the Court they "must be of that quality which indicates an adherence to or a furtherance of the purposes or objectives of the proscribed organization *as distinguished from mere cooperation with it in lawful activities*." [3] The term "sympathetic association," as used in the Loyalty Order, adds something entirely novel in American law. It apparently denotes a lesser degree of organizational connection than is involved in affiliation. No doubt it brings within the range of suspicion the "mere cooperation in lawful activities" which the Supreme Court thought to be inadequate as an evidence of affiliation. Thus, for example, a money contribution for a specific and entirely permissible purpose might reflect a "sympathetic association." As a matter of fact, the term has been given an even more extended significance in the day-to-day work of the loyalty boards. "Sympathetic association" with a proscribed organization has customarily been inferred when one is a relative or friend of another person who in turn has been identified in some way with the organization in question.

In any event, neither membership in nor association with an organization serves, alone, to establish disloyalty. President Truman publicly declared, some months after issuing his order but before its active administration commenced, "Membership in an organization is simply one piece of evidence which may or may not be helpful in arriving at a conclusion as to the action which is to be taken in a particular case." [4] Moreover Chairman Richardson of the Loyalty Review Board has stated that "advocacy of whatever change in the form of government or the economic system of the United States, or both, however far-reaching such change may be, is not disloyalty, unless that advocacy is coupled with the advocacy or approval, either singly or in concert with others, of the use of unconstitutional means to effect such change." Hence, he concluded, "all employees, and all who may aspire to become employees, of the Government, should not only be, but feel, free to join, affiliate or associate with, support or oppose any organization, liberal or conservative, which is not disloyal." [5]

This remark serves to stress what is one of the central problems in the loyalty program. A man may be deemed disloyal if he has associated with a "disloyal" organization as distinct from one which is merely "liberal or conservative." Obviously, therefore, great importance attaches to the choice of the adjective that may best describe a particular group. The Loyalty Order vests that choice in the Attorney General. By virtue of the Order he must pass upon the characteristics of all organizations; and when he has done so, urgent consequences at once appear.

The Attorney General's Black List

The Order contemplates that the loyalty boards will receive from the Attorney General a list of the organizations which, "after appropriate investigation and determination," he has designated as totalitarian, fascist, communist, subversive, or committed to the use of violent or unconstitutional methods. These, it may be supposed, are the disloyal organizations to which Chairman Richardson made reference. Once the Attorney General has spoken, his conclusion is incontestable before the loyalty boards; neither the listed organization nor the employee who has had sympathetic association with it enjoys the privilege of trying to show that the Attorney General was mistaken.

During the administration of Attorney General Clark nearly two hundred groups were identified by him as coming within the scope of the Order. The task cannot have been easy. The terms used in the Order have no well-defined meaning, either in dictionaries or in common parlance. They can be made to mean pretty much whatever one may choose. As recently as February 3, 1949, for example, Senator Taft assured the National Federation of Women's Republican Clubs that "the fundamental cleavage" between the Republican and Democratic parties was "free government versus totalitarian government," [6] an application of the word "totalitarian" which the Truman Administration would scarcely endorse. It is interesting to know, too, that only two decades ago when the responsibilities of national administration were borne by Presidents Coolidge and Hoover, the board of trustees of the American Medical Association denounced as "communistic" the provision of publicly supported medical care for veterans whose illness was not directly connected with their military service.[7] In 1947 President Truman asserted in a formal message to Congress that the real estate lobby had engaged in "subversive" activities in seeking to terminate rent control.[8] The amorphous character of such words as "fascist" and "communist" has long been familiar. In some circles the American Legion, the Daughters of the American Revolution, and the

National Association of Manufacturers are regularly characterized as "fascist," while it is well known that the Interstate Commerce Commission, the income tax, and workmen's compensation have in their respective days shared the quality of being deemed "communist." The Attorney General must have been puzzled to know how to draw the lines which would help him gauge an organization's dominant characteristic. The definitions that the Department of Justice created have never been divulged, but they are perhaps deducible from the listings themselves.

All but one of the twenty-two "totalitarian" organizations on the black list compiled by the Attorney General were connected with prewar Japan—the Black Dragon Society, the Hinode Kai (Imperial Japanese Reservists), and so on; the one exception is the Peace Movement of Ethiopia, an organization among Negroes which in the early days of the war sought to stress the common interests of the world's colored populations. If the Attorney General has accurately sensed the meaning of "totalitarian," it seems reasonably clear that the Government is not shot through with so many totalitarian influences as ex-President Hoover thought it was when, at the Republican National Convention in 1948, he wholeheartedly attacked the "totalitarian liberals" and the "totalitarian economics" of the New Deal.[9] It may be assumed that few sympathizers with Japanese imperialism remain in the federal service today.

The "fascist" organizations, as identified by the Attorney General, also number but twenty-two. Nine are relics of the Nazis, such as the Ausland-Organization der NSDAP and the Kyffhaeuser Bund. Four are reminders of the departed glories of Mussolini, such as the Lictor Society (Italian Black Shirts). The remaining nine have American names—ranging from American Patriots, Inc., to National Blue Star Mothers of America. The initial listing focused on organizations that were linked with the defeated regimes of Italy or Germany. The guiding principle of selection was thus plain. The addition of American organizations through a supplemental list leaves the definition less certain. There is a separate grouping of eight organizations that are said to have "adopted a policy of advocating or approving the commission of acts of force and violence to deny others their rights under the Constitution of the United States"—the Ku Klux Klan, the Silver Shirt Legion of America, and others. The distinction between the two, those which are violent and those which are fascist, is not brought out.

The number of "subversive" organizations is reassuringly small according to the list the Attorney General has compiled. There is the German-American Bund, now extinct; the Communist Party, U.S.A., along with the Communist Political Association, its former alter ego, and the Young Communist League, its wholly owned subsidiary; the Socialist Workers Party; and the Workers Party. In other words, there are only three present-day subversive groups, each of which professes to be the true exponent of Marxism and two of which are markedly anti-Stalin in orientation. This listing suggests, though it does not explicitly declare, that a "subversive" organization is one which teaches that the Government must ultimately be overthrown by violence in order to achieve a new economic order.

Still another listing names the organizations that "seek to alter the form of government of the United States by unconstitutional means." One might have supposed that this list and the roster of "subversive" groups would be coextensive. But they are not. The German-American Bund, while "subversive," apparently believed in the Constitution, for it is not on the "unconstitutional" list. On the other hand, the Industrial Workers of the World and the Nationalist Party of Puerto Rico are not "subversive," but are said to favor unconstitutional methods. The three rivals for the mantle of Marx have been awarded places on both lists.

Now we come to the real meat of the black list. There are 108 "communist" organizations. This is a very confusing grouping, in terms of a deducible definition. The Communist Party and the Young Communist League, which are "subversive" and seekers of the unconstitutional, are also "communist." But the "communist" groups include over a hundred that are neither "subversive" nor in favor of unconstitutional methods. The range is noteworthy. The American Committee for Protection of Foreign Born rubs shoulders with the Council on African Affairs, Commonwealth College, and the Washington Bookshop Association. If for the moment we put to one side those groups that have been separately named as "subversive," the striking characteristics of the "communist" organizations seem to be these:

1. Their ostensible purposes are without exception legal and, according to one's taste in these matters, at least debatably laudable;
2. They have numbered Communists among their active supporters or officers, which gives rise to the suspicion that they may have purposes in addition to or even different from those they avow.

The second of these characteristics deserves slightly expanded attention. What we are saying is that the apparent purposes of an organization may attract many non-Communists, but that if Communists are able to exercise influence in the organization, they are likely to divert its energies into other channels. Thus, for example, many a non-Communist might join the International Workers Order (which is on the black list) because as a legally authorized insurance company it sells small policies at advantageous rates, while at the same time it affords its members various cultural and social opportunities. But there is always the possibility that the Communists who staff the International Workers Order may seek to enlist the organization or its non-Communist members in pro-Communist acts of an entirely political aspect. Clearly there is nothing that is distinctly Communist about the business of writing life insurance or the promotion of Old World folk dancing. The chief differentiation between the IWO and the Metropolitan Life Insurance Company or the Gaelic Society of New York is that Communists are strategically placed in the former and not in the latter two.

Of course this attempt to reconstruct the applicable definitions may not have succeeded. Whatever the definitions may be, they are made operative without even notifying the affected groups that they are under scrutiny and without hearings or informal conferences in which there would be opportunity to establish the character of an organization or its sponsors prior to denunciation by the Attorney General. Mr. Justice Clark, when he was Attorney General, told me that the decision to black-list was never lightly made. He asserted that after subordinate attorneys had analyzed an FBI report concerning a suspect organization, each of his chief assistants was called upon to review a recommended decision; and in cases where opinions were divided, the matter was studied by the Attorney General himself. No doubt the problem is approached soberly, as indeed it should be in view of the effects an adverse decision may have upon an organization's members and upon its own future. Black-listing inevitably causes a decrease in membership rolls, a reduction in contributions, a loss of status as a tax-exempt organization, and considerable harassment in the form of interference with meetings and the like. Whether or not these consequences give rise to questions of constitutionality, as some authorities believe,[10] they are certainly too serious to permit incautious exercise of the great discretion the Attorney General possesses.

As previously noted, the Attorney General's designation is conclusive and may not be disturbed—unless the courts should at some time in the future manifest a hitherto unrevealed readiness to do so. At any rate, the loyalty boards are forbidden to receive from an affected employee any evidence he might wish to proffer concerning the true character of a black-listed group with which he had been linked. This seems to be procedurally sensible. If each loyalty board were compelled to admit proofs and argument about the soundness of the Attorney General's judgment, there would be great duplication of effort and, in time, conflicting determinations. But while there may be wisdom in barring an individual from challenging the black list, despite its important bearing on his own future, the same argument cannot be made against the Attorney General's granting a full hearing to the organization itself in advance of denouncing it.

Some, though certainly not all, of the Attorney General's ex parte conclusions are debatable ones, even if the supposed premises are accepted without challenge. For example, the North American Committee to Aid Spanish Democracy was included in a supplemental list of "communist organizations," though it has been defunct for a full ten years. This immediately elicited a public protest by James Loeb, Jr., national executive secretary of Americans for Democratic Action, and Roger N. Baldwin, then the director of the American Civil Liberties Union, who said in a letter to the Attorney General:

"We write as executive officers of two organizations whose undeviating opposition to Communist and Communist-front organizations is generally known and recognized . . . We were both actively associated with the North American Committee to Aid Spanish Democracy as members of the Executive Committee of that organization . . . During the Spanish War, 1936–1939, it was the only broad national group which aided the legitimate and duly-recognized Republican Government of Spain . . . Conclusive proof of the control of the Committee came in the period following the Nazi-Soviet Pact and the outbreak of the war later in 1939. The pro-Communist elements were defeated in their efforts to use the Committee to embarrass the Western Allies then at war against Germany, Soviet Russia's ally. The pro-Communist elements were forced to withdraw from the Committee and establish their own group . . . Meanwhile, the Spanish Refugee Relief Campaign, a direct continuation of the North American Committee, continued to operate for several years, giving assistance to non-Communist Spanish Republican refugees. . . ."

Or, to suggest a more contemporary example, consider the case of the American Russian Institute, located in New York and unconnected with similarly named organizations elsewhere in the country. An early list of "communist organizations" contained a reference to the "American Russian Institute." When the directors of the American Russian Institute called informally upon the Attorney General to protest the listing, he acknowledged that a mistake had been made; he informed the Loyalty Review Board that the adverse listing should be confined to "American Russian Institute (*of San Francisco*)," which was functionally and organizationally distinct from the more widely known Institute in New York.

So the matter stood until April 21, 1949, when, without any prior indication of a change of mind, Attorney General Clark notified the Loyalty Review Board that he had added to his black list the following among other organizations:

"American Russian Institute, New York
"American Russian Institute, Philadelphia
"American Russian Institute of Southern California, Los Angeles."

The American Russian Institute of New York, organized in 1926, describes itself as "nonpolitical and nonpartisan, devoted to research on the Soviet Union, and dissemination of the results." It maintains a large reference library, open to the public; publishes a magazine, the *American Review on the Soviet Union,* and an indexed digest, *Russian Technical Research News,* which makes Soviet technological information available to commerce, industry, science, and the Government; and provides translating and research services for newspapers, business firms seeking to trade with the USSR, writers, and students. Its active directors number among other prominent citizens William W. Lancaster, a senior partner in the distinguished law firm of Shearman & Sterling & Wright, counsel to the National City Bank, and Richard B. Scandrett, Jr., a leading New York Republican who has been vigorous among the supporters of Senator Taft. When the directors of the American Russian Institute once more visited Attorney General Clark to tell him that he had made an erroneous classification, he reportedly replied that when he listened to them, his listing of the organization as "communist" seemed perfectly silly, but that when he listened to his assistants, they assured him that the listing was correct.

These illustrative comments about two organizations fall far short of establishing that Attorney General Clark was mistaken in his characterization of them or of any others. In all probability the Attorney General had received confidential information about these groups which made him suspicious of their nature. Even in these days of hypersensitivity it is unlikely that an organization would be denominated Communist merely because it believed that what happened in one-sixth of the world, the Soviet Union, was a matter of legitimate intellectual interest in the United States. This much, however, is clear. The Attorney General's possession of additional confidential information does not establish the correctness of his findings. Confidential information, like any other, may be incorrect or misleading because incomplete. When so much of what appears on the surface of an organization is unexceptionable, there is a serious enough question to warrant an orderly inquiry before concluding that the surface is a sham. Thus far no procedure for making such an inquiry has been provided. The issues are too important to be left to intuitive judgments or to untested appraisals of possibly imperfect evidence.

In one other respect the Attorney General's black list seems markedly unfair to those against whom it may be used. Anyone who has been exposed to political realities in the past twenty years knows that many organizations have changed their orientation during that period. The black list reflects no appreciation of this commonplace of American life.[11] An organization with entirely lawful purposes may at some time have been "captured" by the Communists. Its name will then appear on the list as though it had been Communist-inspired and Communist-controlled from its inception, and all who have had contact with it at any stage are thereupon tainted. In only one instance in the whole long black list is there mention of a date that shows when a previously unobjectionable organization became sufficiently Communist to warrant its being denounced. That single exception is "Nature Friends of America (since 1935)." Until 1935, apparently, but not afterward, one could associate with that obscure group guilelessly and without a qualm, motivated solely by friendliness to nature.

All in all, the black list is a rather blunt instrument to use in probing the subtleties of motivation and beliefs which bear on loyalty.

The Discovery of Disloyalty

This excursion into the black list's meaning and method has been necessary because so much of the quest for disloyalty revolves about that catalog. Note that the object of the search is disloyalty rather than loyalty; or perhaps it would be more accurate to say, as has President Truman, that the search is for the "potentially disloyal." [12] Nowhere in the Loyalty Order or in the directives, pronouncements, and judgments that have grown out of it has an effort been made to isolate loyalty as an affirmative quality. All the standards contained in the Order are suggested for use as tests of the possible absence of loyalty, rather than as means of discovering its presence. Responsible administrators have been asked in numerous personal interviews to describe the determinants of loyalty. Some have responded that the answer was self-evident. Others have explained their measuring rods of *dis*loyalty. Not one has put the matter positively.

Their failure to do so is understandable. In the glowing words of the distinguished historian Henry Steele Commager, loyalty "is a tradition, an ideal, and a principle. It is a willingness to subordinate every private advantage for the larger good. It is an appreciation of the rich and diverse contributions that can come from the most varied sources. It is allegiance to the traditions that have guided our greatest statesmen and inspired our most eloquent poets—the traditions of freedom, equality, democracy, tolerance; the tradition of the higher law, of experimentation, and of pluralism. It is a realization that America was born of revolt, flourished on dissent, became great through experimentation." [13] To measure men against so high a standard of idealism might produce too many failures. It is no doubt safer and wiser to employ the somewhat less stringent because much narrower negative tests that the loyalty administrators have announced.

What, in brief, are those tests? As has already been seen, mere identification with a black-listed organization is not conclusive proof of disloyalty, though membership in the Communist Party or a splinter of it is an independent cause for dismissal under the Hatch Act and the appropriations statutes. Nor, so the Loyalty Review Board has announced, is disloyalty evidenced by "advocacy of whatever change in the form of government or the economic system of the United States, or both, however far-reaching such change may be," so long as unconstitutional means of effecting the changes are not advocated. In summarizing its policies, the Board has identified only two classifications of persons who should be disqualified from federal service:

1. "Persons holding beliefs calling for a change in our form of government through the use of force or other unconstitutional means, who indicate these beliefs by association or conduct"; and
2. "Persons who demonstrate that their allegiance is primarily to some foreign power or influence, and that they desire to overthrow our Government." [14]

These tests seem fairly precise. One cannot quarrel with them as abstractions. Of course existing laws bar the first group, that is, those who seek to alter our governmental structure by force. The Loyalty Order, under the announced interpretation of the Loyalty Review Board, adds nothing in that respect. As for those who have demonstrated their allegiance to a foreign power or influence, looking toward overthrow of our government, elementary principles of self-defense support ousting them from posts of power.

Unfortunately, however, there is little seeming correspondence between the announced tests and the actual administration of the Loyalty Program. Let us study a few of the "charges" and questions that have been deemed to bear on the issue of loyalty.

A recent case involved the fate of a young psychologist employed by the Veterans Administration at one of its hospitals. The first charge against him reads as follows:

"That you were a member of the American Labor Party of New York, New York, in 1938 and 1939, which was cited as a Communist front organization by the Committee on Un-American Affairs in 1946."

The Loyalty Order, it will be recalled, speaks of the significance of membership in an organization that the Attorney General has designated as "communist" or otherwise improper. Here, as in many other cases which have been studied, may be seen a broadening of the ranks to admit organizations which others than the Attorney General have denounced. The Attorney General has listed 108 "communist" organizations. The House Committee on Un-American Activities has significantly bettered this record. On December 18, 1948, it published a list of 564 organizations and 190 publications "which have been declared to be outright Communist or Communist-front enterprises." [15] Groups that the House Committee has stigmatized, such as the Southern Conference for Human Welfare and the United Public Workers of America, are often cited in loyalty proceedings, although the Attorney General has made no adverse determination concerning them. Consequently, a wary federal employee cannot be content to check his associations against the Attorney General's black list; other lists may yet confound him. The safe thing is to shun all associations whatsoever.

In the present case even that would not have been enough. The defendant was separately charged as follows:

"That in 1941 your name was on the active mailing list of the American Spanish Aid Committee, an organization controlled by the Communist Party."

This organization does not appear on the Attorney General's list; the source of information that it was "controlled by the Communist Party" is not indicated. But even if that control

146

did exist, the defendant here was charged with no organizational connection other than that his name appeared on a mailing list. If it did so appear, he testified, it had been placed there without his request or consent. Anyone who reviews his own incoming second-class mail over a period of time must find himself bemused and baffled by the variety of "sucker lists" that inexplicably contain his name and address.

As for the merits of the charge concerning membership in the American Labor Party in 1938 and 1939, the dates are of interest. In 1938 our young psychologist had just graduated from college, at the age of twenty-two. When he joined the American Labor Party in that year it was the party of La Guardia and of many idealistic New Yorkers who found slight comfort in the local organizations of the Democrats and Republicans. In 1937 it had helped return Mayor La Guardia to office; in 1938 its major candidate was Governor Lehman, a candidate for re-election. In 1939 the ALP supported President Roosevelt's foreign policy, then under Communist attack, and condemned Communists as "betrayers of the labor movement, antagonists of democracy, and protagonists of dictatorship." The "right wing" firmly controlled the party's offices and policies. In 1940, as in 1939, the ALP's platform endorsed the national defense program and denounced the "tools" of the Nazi-Soviet pact, which then existed. Not until later years and many vicissitudes, which need not now be detailed, did the American Labor Party burst asunder, with the formation of a new Liberal Party under the leadership of David Dubinsky and abandonment of the ALP to his political opponents, including Communists and their supporters. This, then, was the lawful and open political party to which a youthful college graduate belonged in 1938 and 1939, only to discover a full decade later that his professional career was shadowed by the retroactive significance of intervening events in which he was not accused of having played any part.

An additional charge against this same unhappy individuaɪ reads as follows:

> "That in 1945 ＿＿＿＿＿＿＿＿ was an officer of the ＿＿＿＿＿＿＿＿ Club of the American Youth for Democracy and that in 1947 you married her."

The American Youth for Democracy has, it is true, been designated by the Attorney General as Communist, because it is the lineal descendant of the Young Communist League. Note, however, that the quoted charge does not assert an association between the defendant and the organization. It asserts, rather, an association between him and an individual who, at an earlier date, had been a member of the black-listed group. According to the record of the hearing in this case, the allegation concerning the young lady may not have been correct. She testified that she had never belonged to, let alone been an officer of, the group in question. Furthermore, she did not even meet her future husband until two years later, so that there could be no question of his having had any influence upon her alleged membership. But even if we accept the charge's assertion at face value, we nevertheless see how tenuous becomes the thread of inference when it is stretched as far as it has been here.

This is by no means an exceptional instance of projecting the concept of sympathetic association beyond the limits of the Loyalty Order itself. A young medical scientist, for example, has been embroiled in difficulties not because of anything she had done in her scrupulously nonpolitical life, but because her father (with whom she was no doubt "sympathetically associated" even though their residences were separated by nearly a thousand miles) was the director of a black-listed organization. A similar embarrassment was visited upon an executive whose aged father has for decades been the recipient of international honors and scholarly recognition; in order to absolve herself of blame, she faced the necessity of establishing not her own but her parent's political purity. In one widely discussed case a distinguished scientist was taxed with having been the sympathetic associate of a member of the I.W.W.; evidence at the hearing established that this dangerous fellow had been a neighbor of the scientist when he was a lad of eight and had then commenced a friendship that could scarcely have reflected political convictions. The reductio ad absurdum came when one of this country's outstanding men of science was challenged because he had assertedly failed to admonish his wife at a private dinner party during which she had made statements an anonymous informant thought were critical of American foreign policy and favorable to the Soviet Union.

These instances suggest that disloyalty may be deduced if the affected employee has had family, friendly, or uxorial relations with any person or persons who might be regarded as possibly disloyal. They serve as a cautionary reminder that a man may indeed be known by the company he keeps, rather than by what he himself does. The logical fallacy in this sort of charge is readily apparent. It confuses personal association with political advocacy or endorsement. It proceeds on the theory than an individual who knows a Communist sympathizer is probably a Communist sympathizer himself, although he may know rock-ribbed Republicans or Dixiecrats equally well without being assumed to be their political confederate. It supposes in effect that if a man has talked with a Communist or has read a Communist publication, he will have been persuaded by everything he heard or read in that quarter, while his more numerous contacts with non-Communists and his avid reading of the Luce magazines will have left him untouched. On the whole this attributes to Communist spokesmen a considerably greater force of personality and persuasiveness than most observers have been able to discover at first hand.

Yet another of the charges against our beleaguered psychologist is interesting. It reads as follows:

> "That in 1941 you interceded with a public official, the Secretary of Labor, on behalf of a known Communist who had been dismissed from public office."

The case in question involved a Mrs. Miller, who was ousted from her post in the Department of Labor by Secretary Perkins on August 1, 1941. The matter arose under the Hatch Act, though for technical reasons the case was brought under the "efficiency of the service" clause of the removal statute. The Department was able to show that the employee had adhered to Communist Party positions in her union and elsewhere, and had solicited a fellow-employee to become a Communist. Since this was the first publicized instance of a civil servant's being discharged because of Communist affiliations, the case attracted considerable attention at the time. The proceedings were in truth carefully conducted, on the basis of fully stated charges and with close attention to the employee's rights. Nevertheless, a prominent union of federal employees urged its members, of whom our psychologist was then one, to protest to the Secretary of Labor against what it characterized as an unfair decision because it had not been preceded by a proper hearing. Apparently without personal investigation, he complied with his union's request, not, it seems, on the ground that the ousted official should not be dismissed even though a Communist, but on the ground that the procedure was improper. Such a communication to a public official scarcely establishes a desire to overthrow the Government.

The significance of the present charge goes well beyond the facts of the particular case. An ominous belief is abroad in the country that really loyal citizens should acquiesce in what they believe to be injustice if the victim of the injustice chances to be a Communist. Thus, for example, the California Committee on Un-American Activities has reported its conviction that the American Civil Liberties Union is a Communist organization because it insists that the protections of the Constitution extend to all alike, even including the Communists.[16] There is nothing particularly novel about this sort of prejudicial identification. Many of the victims of delusions about witchcraft in Salem in 1692 were not, as legend has it, the misshapen and unloved crones of the community, but were respectable people who sought to withstand the mania; for their efforts they were promptly "cried out upon" as being witches themselves.[17] In our day an academic scientist, who had agreed to do a job for one of the military services for a period of six months, was charged with disloyalty before finishing the job because "During your period of employment at the University of ——— you made statements to the effect that you believe 'the House Un-American Activities hearings in Washington, D.C., are more of a threat to civil liberties than is the Communist Party because they infringe upon free speech and if this sort of thing is continued there is more danger of fascism in this country than communism.' Further, you have argued that 'as long as the Communist Party is legal it is the duty of every one to protect the Party's rights.' "

The fact that there is nothing novel about ascribing base motives to dissenters does not render any more desirable the present tendency to discourage conscientious protest by identifying disagreement with disloyalty. The loyalty boards have in many cases reflected this tendency by closely questioning defendants concerning their attitude toward the Loyalty Order itself, thus perhaps stimulating a widespread readiness to "crook the pregnant hinges of the knee where thrift may follow fawning." If a man acknowledges belief that the Loyalty

Order does not contribute to the growth of American democracy, this belief may in itself induce a finding that he is disloyal. Questions with equally clear implications have frequently been asked concerning a man's opinions about the Marshall Plan, or American influence in Italian elections, or world federalism. Federal employees have even been interrogated about possessing Paul Robeson records or Howard Fast novels, as though to suggest that artistic commendability and political eligibility are as closely linked in this country as they appear to be in Russia.

The passion for conformity is still more seriously manifested in the field of civil liberties in general and race relations in particular. The chairman of a departmental loyalty board, an amiable and devoted public servant, said to me one day, "Of course, the fact that a person believes in racial equality doesn't *prove* that he's a Communist, but it certainly makes you look twice, doesn't it? You can't get away from the fact that racial equality is part of the Communist line." It comes as no great surprise, therefore, to learn that in a proceeding involving a scientist who had actively participated in the wartime development of the proximity fuze, a member of this loyalty board asked the scientist's supervisor:

"Have you had conversations with him that would lead you to believe he is rather advanced in his thinking on racial matters?—discrimination, non-segregation of races, greater rights for Negroes, and so forth?"

In a case in a different agency a highly rated professional employee was summoned to defend himself against the following charges which were deemed to bear on the issue of disloyalty:

"A confidential informant, stated to be of established reliability, who is acquainted with and who has associated with many known and admitted Communists, is reported to have advised as of May, 1948 that the informant was present when the employee was engaged in conversation with other individuals at which time the employee advocated the Communist Party line, *such as favoring peace and civil liberties* when those subjects were being advocated by the Communist Party.

"Another informant, reported to have been acquainted with the employee for a period of approximately three years, from 1944 to 1947, reportedly advised that while informant did not have any concrete or specific pertinent information reflecting adversely on the employee's loyalty, informant is of the opinion *that employee's convictions concerning equal rights for all races and classes extend slightly beyond the normal feelings of the average individual,* and for this reason informant would be reluctant to vouch for the employee's loyalty."

It is a wry commentary on the loyalty program that charges like these supply an official endorsement of the Communist Party's propaganda line. The Communists proclaim themselves to be firm believers in peace, civil liberties, and human decency. It seems to many steadfastly democratic Americans that they, rather than the Communists, ought to be given major credit for these laudable beliefs.

In the particular proceeding under discussion there occurred an exchange of questions and answers that encouragingly illustrates the survival of a free soul under pressure. The employee's former superior was called as a witness, and under questioning testified as follows:

"Q. Getting back to the question of civil liberties, would you say that his feelings about racial relations were of such a nature to indicate to you that he was a member of the Communist Party?

"A. Mr. ———'s opinions on the matter of racial relations

are that he was strongly in favor of equality and equal
rights for Negroes. If that makes a man a member of the
Communist Party—why, I suppose it makes me one,
and I think it probably makes some of you gentlemen
[members of the Loyalty Board] one.

"Q. [By the Board chairman] Would you say that Mr.
————'s activities concerning civil liberties were no
greater than that of the average American person?

"A. No, I would say that his interest in civil liberties was
certainly greater than that of the average. I think it
is very unfortunate that the average American is not
sufficiently interested in civil liberties except when his
own are affected. He can get pretty hot about his own,
but in too many cases he just isn't strongly interested
in what happens to other people, particularly people
of different groups."

Not all would speak so courageously at a time when a torpid
social conscience is a strong guarantor of security and comfort. It is well and good to say that everyone should have the
courage of his convictions. But few people do in fact have
the fortitude to cling to beliefs that may expose them to
calumny and loss. One of the virtues of democracy is its
maintenance of a climate in which normally timid persons
are allowed to entertain opinions without having to demonstrate heroic qualities. The central tenet of the democratic
philosophy is that governmental policy should be shaped by
the discussion of men who are free—free to inquire, to compare, to experiment, to debate, and to complain. The loyalty
program drifts in the direction of curtailing that freedom.

Consider, in terms of its implications for democracy, the
case against a former university professor who had served for
more than five years in an important post and who had re-

ceived the Certificate of Merit for his share in studies that were put to immediate strategic use during the war. He was told that information was at hand showing that he had "participated in Communistic activities" and had "exhibited a proCommunist and pro-Soviet attitude for several years"; in support of these generalized statements the following specifications appeared among others:

> "That you protested the dismissal of a teacher for his Communist teachings; that you favored resolutions to free Tom Mooney . . ."

A former governor of California, a Democrat like the defendant, came to his aid, saying, "If favoring resolutions to free Tom Mooney is evidence supporting a charge of disloyalty, then it applies to millions of disloyal Americans throughout the nation including Republicans, Democrats, conservatives, liberals, business men and workers, bankers and lawyers, members of the American bar who investigated the case, and myself who, as Governor of California, pardoned Tom Mooney." The allegedly Communist teacher whose ouster the defendant had opposed was, according to testimony at the hearing, dismissed because he appeared on a picket line during a lumber strike; the action was said to have been opposed by a very large number of California teachers because of belief that an issue of academic freedom was involved.

Speaking in his own behalf, the defendant said in closing his case: "I have a great personal stake in America. However, I feel it is the duty of every citizen in a republic to participate to the extent of his desire to strengthen it. Particularly, educated people have a special responsibility to use their training for the public good. I have worked both as an individual and a part of a group in strengthening things I believe in. I have tried to determine my stand on issues on the basis of merit, without waiting to determine whether the communists would be for it or against it."

In the end this particular individual was "cleared." As a matter of fact most of the persons who face the terror, shame, and expense of answering charges of disloyalty are finally acquitted. The latest available figures, as of May 31, 1950, show that of all the cases that went to hearing, less than 13 percent resulted in finally adverse determinations; and if to this is added all these whose cases were still in process of appeal or reconsideration, the total rises to only about 19 percent. This may be contrasted with the percentage of convictions obtained in ordinary criminal cases that are brought into federal and state courts after the return of an indictment. In New York County, over 98 percent of all persons who were indicted during 1946, 1947, and 1948 either pleaded guilty or were convicted after trial; of the cases that actually went to trial, over 85 percent resulted in conviction or admissions of guilt after the evidence had been presented. In the federal courts over the same three-year span 84 percent of all defendants against criminal charges were convicted or pleaded guilty. The striking disparity between these records and the record of the loyalty boards is not a reflection of different attitudes upon the part of the judges. Trials in the federal courts and in those of New York are notably fair; there is no inhumane disposition to hold the innocent guilty. The difference is that loyalty boards commence proceedings against federal employees, involving the scandalous imputation of disloyalty and jeopardizing their whole careers, on far flimsier evidence than will move a prosecutor to proceed against a pickpocket or a stock swindler. In large part this reflects the Loyalty Review Board's conception of the function of hearings. The Board has told the subordinate loyalty boards that charges and hearings are to be deemed merely a part of the process of investigation. Hence the loyalty boards have been urged to issue charges whenever their files contain any unexplained "derogatory information," even when that information is on its face inadequate to sustain a reasonable belief that the employee may be disloyal. A memorandum from the Loyalty Review Board emphasizes that "unless the Board concludes from an examination of the whole record that an employee is *clearly eligible,* it is desired that the Board proceed to dispose of the case after hearing and not by determination without hearing." [18] Largely because of this instruction, cases may go to trial for very flimsy reasons and without any real expectation that a finding of disloyalty will be made, for as we have seen "derogatory information" embraces everything that suggests even a rather remote relationship with an objectionable organization or an individual.

Social Results of the Loyalty Program

If the only effect of this were upon the individuals who suffered the costs and concern of facing loyalty charges, the matter would be serious. The shattering financial, psychological, and practical consequences of even a wholly successful defense against charges are commonly recognized, for the stigma is not erased by a clearance and nothing can replace the harrowing months of uncertainty and the loss of friendships that are usual concomitants of loyalty proceedings. If, however, this suffering were offset by an important gain, we might then perhaps be able to agree that efforts to enhance the tone, quality, and reliability of our civil servants warrant the incidental and unmalicious destruction of a few of them.

But there is more at stake than this.

In the field of science, the crudities of the loyalty program discourage efforts to draw into public service the live-minded and experienced men whose talents are needed in many agencies. The distress occasioned by an unwarranted inquisition by a loyalty board is felt by a wide circle of friends and fellow-workers. Especially in the case of scientists there is a realization that even after a man has been exonerated following a hearing, he may still be subjected to a renewal of the charges and a dusting off of the same evidence if the winds of politics continue to blow strongly. On September 6, 1948, eight of America's great scientists, joining in a message to President Truman and Governor Dewey, deplored the disastrous effects upon scientific recruitment that followed the denunciatory sensationalism of the House Committee on Un-American Activities. Harrison Brown, professor of nuclear chemistry at the University of Chicago; Karl T. Compton, then the president of Massachusetts Institute of Technology; Thorfin R. Hogness, director of the Institute of Radiobiology and Biophysics at Chicago; Charles C. Lauritsen, professor of physics at California Institute of Technology; Philip McC. Morse, then professor of physics at M.I.T. and now operations director of the Weapons Evaluation board under the Secretary of National Defense; George B. Pegram, vice president of Columbia University; John C. Warner, dean of the graduate school of Carnegie Institute of Technology; and Harold C. Urey, professor of nuclear physics at Chicago, concluded that the atmosphere of suspicion surrounding scientists in government was an effective deterrent to procurement and use of their services. What these men said publicly has been echoed privately by scientific men of every level of eminence.

The negative consequences of the Loyalty Order are dramatically realized when able men refuse to engage in public service or choose to leave it for less harassing occupations. All in all, however, the more serious though perhaps more subtle impact is on those who remain in federal service. Former Attorney General Clark remarked in my presence in June of 1949 that never before had the morale of federal officials been so high, thanks to the Loyalty Order. Numerous conversations with Government employees have led me to a completely contrary conclusion. Time after time there has been a reflection of suspicion and reserve in human relationships both within and outside the ranks of fellow-employees. One small anecdote is illustrative of many. During 1949 a young scientist was on leave from a federal department in order to complete his graduate work at Columbia University. While he was in residence in New York, his landlord made application to the federal rent control authorities for permission to increase the rent on his apartment by 30 per cent. Other occupants of the large apartment building in which he lived, being faced with the same threat of higher housing costs, requested the help of a neighborhood Tenants Council. This membership group, which employs trained investigators and attorneys, represents tenants who might individually be unable to resist unwarranted rent increases. The young federal scientist, threatened by a formal proceeding in which neither his funds nor his available time would permit his participation, desired to turn over his case to the Tenants Council, as he could do by becoming a member and paying monthly dues of fifty cents. Before doing so, however, he asked a Columbia professor to inquire into the political orientation of the organization. According to the information he received, the Tenants Council engaged legitimately and with reasonable success in its declared work of opposing improper rental demands. But the group was said to have been inspired by and to be largely under the continuing administrative control of members of the American Labor Party. The American Labor Party, in turn, has in late years been heavily infiltrated by Communist elements, so that, though it is not entirely Communist, it is no longer, as it once bade fair to be, the chief political vehicle of organized labor in New York City. Upon being told these facts, the youthful federal employee sighed and said: "Well, I'll just have to try to survive the rent increase. I certainly can't go to the expense and trouble of fighting it personally. And I'm afraid I can't afford to risk a membership in the Tenants Council. After I receive my doctorate I'm planning to return to the Department. And with things as they are in Washington, one can't be too careful."

It is in the unrecorded accumulation of undramatic episodes like this one that the true effect of the Loyalty Order can be discerned. It has not unmasked spies and saboteurs—indeed, Chairman Richardson of the Loyalty Review Board recently told the Senate that "not one single case of espionage" had been encountered during the three years of the loyalty program, and that the FBI had found no evidence even "directing toward espionage" in the course of its 10,000 full field investigations and 3,000,000 examinations of records.[19] The Order has not led to the discovery and ouster of hordes of Communists. It has not, as Mr. Clark asserted it had, encouraged tranquillity of spirit among federal employees. What it has done—and perhaps not even designedly—is to enforce a new concept of loyalty. This "new loyalty," as Professor Commager has summarized it, "is, above all, conformity. It is the uncritical and unquestioning acceptance of America as it is—the political institutions, the social relationships, the economic practices. It rejects inquiry into the race question or socialized medicine, or public housing, or into the wisdom or validity of our foreign policy. It regards as particularly heinous any challenge to what is called 'the system of private enterprise,' identifying that system with Americanism. It abandons evolution, it repudiates the once popular concept of progress, and regards America as a finished product, perfect and complete."[20]

Some of the cases involving a federal scientist have included charges that correspondence or contact had been had with some other scientist whose politics were unacceptable. The possibility that a professional acquaintanceship may lead to the opprobrium of a loyalty hearing does not encourage Government employees to cast themselves incautiously into invigorating currents; the fermenting ideas of science do not always arise in the minds of individuals who can survive ideological litmus paper tests. Nor is zeal for scientific inquiry engendered by fear of political embarrassment. The relationship between the Loyalty Order and professional freedom is well illustrated by the following chronicle.

Commencing in 1943 the American-Soviet Medical Society published in this country the *American Review of Soviet Medicine,* a journal which, as its name suggests, contained translations of articles and reports which had originally appeared in Soviet medical periodicals. Previously the Soviet medical literature had been unavailable in this country, partly because of the language barrier and partly because of the difficulty of obtaining Soviet publications. The *American Review of Soviet Medicine* therefore importantly contributed to our country's knowledge of scientific developments in the Soviet Union, having nothing to do with its economics or with world politics. The October 1943 issue, which was the first, contained translations of articles on "Treatment of Fresh Wounds by Transplantation of Chemically Treated Tissues," "Gunshot Wounds of the Blood Vessels," "Spasokukotski's Method of Feeding in Penetrating Abdominal Wounds," and a number of others that bore on the immediate problems of military surgery. The October 1948 issue of this publication was its last. It contained articles on "The Toxins of Moulds," "Properties of Snake Venom," "Influence of the Spleen on Migration of Ca and Na from Skin and Muscle," "The Influence of Bromide on Castrated Dogs," "The Problem of Scarlet Fever in Public Health Care of Children," and "Fat Embolism in War Trauma Associated with Lesions of Long Bones." The preceding issue had contained articles on "Virus Etiology of Acute Nephritis," "Experimental Phobia," "Surgery for Cancer of the Esophagus," and "Rheumatic Granulomas of the Lung." Earlier numbers had been devoted to current papers dealing with cancer, tuberculosis, and other areas of active research in both the Soviet Union and this country.

Until the issuance of the Loyalty Order, this scientific journal seems to have been widely read by American physicians engaged in work that might be advanced by knowledge of the results reported by Soviet colleagues similarly engaged. At that time, according to information obtained from the periodical's business manager, there were some 600 members of the American-Soviet Medical Society in Washington; after that Order had been in existence for less than two years, the membership had shrunk to thirty. In March of 1947 there were some 150 subscriptions in Bethesda, Maryland, where are located the National Institutes of Health and the United States Naval Hospital; when the magazine suspended publication, not a single one had survived. In the interval, this non-political magazine had received numerous requests that it be mailed in a plain wrapper, not bearing the publication's name. The conclusion is inescapable that insecurity, engendered in significant measure by the Loyalty Order, caused a flight from exposure to a potentially important body of scientific literature. In order to avoid doubt about their loyalty, federal medical scientists appear to have felt that they must remain ignorant of Soviet researches that might very possibly have furthered their own work in American laboratories. An ironical sidelight on this episode is that the American-Soviet Medical Society had experienced mounting difficulty in obtaining Russian publications from which it drew material for American distribution; the Russians, with a xenophobia that very probably exceeds our own, were seemingly reluctant to let Americans have the benefit of the Soviet scientific findings.[21] Interestingly enough, the flow of Soviet medical journals to this country has recently resumed its former dimensions; but there is no longer an American publication that can readily make their contents known to our professional men.

We need not speculate about the possibility of grievous harm to America if there be insistence upon "political correctness" before a scientist may serve his country and his fellow men. German science deteriorated during the Nazi regime not merely because persons of Jewish descent were expelled; they were, after all, only a small part of the scientific population, though as individuals many of them were important figures. Nor was German science brought to its knees by the mythology and pseudo learning which were intended to obscure the errors of racism. True, anthropology and the social sciences were distorted beyond recognition and "Jewish ideas" were tabu in other branches of learning, while "pure research" was frowned upon and "practical" work was encouraged. Even so, genuine scientific effort remained possible. Good work continued to be done in synthetics, rockets, jet propulsion, and other areas. But the previously high quality of research became increasingly spotty. What chiefly sapped the vitality of the German laboratories was that responsibility was too often entrusted only to those who were "politically reliable." The director of all war research in German universities, for example, was Rudolph Mentzel, a second-rate chemist who had risen to be a brigadier general in the Elite Guard. The Army's research program was placed in the charge of a mediocre physicist named Erich Schumann, whose prior studies, at least as reflected in his writings, had been confined to the vibrations of piano strings. Bernhard Rust, Hitler's Minister of Education and a man of no scientific pretensions, was long the overlord of all the state-controlled research institutes. Karl Brandt, Major General in the Nazi Elite Guard, served as Reich Commissioner for Health and Sanitation, and in that capacity superintended a diabolical and ineffectual program of medical research upon living human bodies. Correspondence between scientific merit and Nazi orthodoxy was fortuitous; it was orthodoxy rather than merit that was the first consideration. The able physicist Bothe was ousted from his professorship at Heidelberg, which was turned over to one Wesch, an inferior scientist but an energetic Nazi as Bothe was not. A captured German report concerning the rocket researches at Peenemünde identifies a Dr. Elvers as an especially competent man, but remarks that he "is merely an anti-aircraft sergeant and thus cannot be placed high in this military establishment." As the "purity" and "reliability" of the scientists became more fully assured, the purity and reliability of the scientific work declined.[22]

More recently there has been mounting evidence that a similar process of politicalizing science is going forward "behind the Iron Curtain." In Hungary, for example, leaders have called for opposition to "reactionary, that is, Western orientation of our scientific and cultural life," while stress is placed upon the desirability of basing physics, chemistry, and astronomy on Marxian principles. In the Soviet Union itself, as has been widely reported, the officially approved Michurin-Lysenko theories of genetics have swept aside those who adhere to "Mendelian errors." [23] Heavy attacks have been aimed as well at physiologists, bacteriologists, and physicists, among others, who have fallen under "western influence" and are therefore politically and scientifically suspect.

No pretense is made here at evaluating the contending scientific ideas that are involved in the Russians' debates. For all that a nonscientist can say, some of the Soviet theories may in time be established as sound, while some of the theories of "Western bourgeois idealists" may prove to be mistaken, as scientific theories often are whether they emanate from West or East. The important thing about the present Russian excitements is, however, that the currents of scientific inquiry and thought have been diverted by political considerations. When a scientist's ability, indeed his very standing to continue in state-supported research, is measured by his conformity to political tests, the more exactingly objective tests of science lose their force.

These developments in the Soviet Union seem to be closely related to the cold war between that country and ours, just as is the loyalty program of the United States. Since the middle of 1947, as the Department of State has reported in its excellent review of cultural relations between the two countries,[24] the Soviet government has placed every sort of legal obstacle (backed by the threat of heavy punishment) in the way of contacts between Russians and foreigners; relations with Americans have been insistently represented as "a threat to the well-being of the Soviet state." Four prominent Soviet scientists visited this country in late 1946 to inspect American cancer-research centers. Included in this quartet was Dr. Vasili V. Parin, then secretary of the Soviet Academy of Medical Sciences and a man who impressed many American colleagues by his objectivity, ability, and personal qualities. The State Department summarized the visit as follows: "All the latest scientific developments were shown the group during its visit. By this time, however, the Soviet Government apparently began to look with suspicion upon those having contacts with the free world. Upon his return to Moscow Dr. Parin apparently vanished. Then, possibly as a sequel, the Soviet Minister of Health was shortly thereafter dismissed." It is perhaps significant that Dr. Parin, in an address before the American-Soviet Medical Society in New York in December 1946, had said: "It is obvious that our plan [for medical research] includes practically the same problems as those studied in the U.S.A. It indicates once more that modern science is really international in character, and proves once more the need for scientific interchange." [25] This preceded by only a few months the promulgation by Russia of a new and stringent State Secrets Act, which was followed by a mounting volume of press attack upon scientists who still maintained Western contacts. Scientists who have published articles in foreign periodicals have been stingingly rebuked and in at least one known instance removed from a post of responsibility. Beliefs about the universality of science and the desirability of exchanging knowledge are insistently discouraged by attacking the probity of American and other "Western" scientists, who are represented as espionage agents or the willing tools of monopoly capital, eager to obtain Soviet scientific secrets as an aid to aggressive war. Americans who have wished to confer with their Soviet counterparts, like the world-famous Russian cancer specialists Doctors Roskin, Kluyeva, and their associates, have been refused visas apparently for no reason other than that the political purity of the Soviet scientists would best be assured by ending their contacts with the outside.

In terms of the advancement of knowledge throughout the world the isolation of Soviet science as a result of its being politically infused is of course unfortunate. The major loss, however, falls on the Soviet side of the barrier. The matter was well put in an address on December 8, 1948, before the New York County Lawyers' Association by Mr. Justice Robert H. Jackson, who had been the United States prosecutor of the major German war criminals in the Nurnberg trials: "I agree," he said, "that the iron curtain is regrettable. But I think it is ultimately more disastrous to those it shuts in than to us whom it shuts out. If they want to handicap themselves by closing the Soviet Union's eyes and ears to the actions and thoughts of the Western World, I do not think it strengthens them against us. If they want to send their scientists to Siberia because they do not make the cold facts of science, such as genetics, support Soviet political theories, I condemn it as inhuman; but I don't think it imperils our security. If it is necessary to maintain Kremlin control over the diversified and scattered Russian people by banishing thought and research, art and drama, that is out of step with their politics, we may deplore it; but we need not lose sleep about its endangering us. The Nurnberg evidence is that the seeds of eventual annihilation for Hitler's power were sown when he began burning books, exiling scholars, persecuting scientists and closing down on information."

We are still far from emulating the Soviet or German policies respecting the content of scientific thought. But let us remember that thoughts are not disembodied entities. They exist in the minds of living people. We take a step, a long and dangerous one, toward scientific immobility when we maintain a program that seeks to fetter minds. The administration of the Loyalty Order has in too many instances laid the paralysis of fear upon federal scientists. They know that disloyalty charges may not only jeopardize their present jobs, but may effectually disbar them as well from nongovernmental employment in their specialties. Minds filled with this sort of un-ease are not likely to be boldly creative in the United States any more than in the Soviet Union.

Perhaps the time has come to consider whether the Loyalty Order deserves to be expunged. In the beginning it was an outgrowth of a decade of political pressure. The House Committee on Un-American Activities set the fire under the pot and kept it roaring from 1938 onward, repetitively proclaiming that the New Deal Administration was shot through with Communists and subversives. In the 1946 Congressional elections the pot was brought to a brisk boil by the Republicans, who placed heavy emphasis upon the need of a large-scale house cleaning in Washington. By 1947 the pot had nearly boiled over completely. The Republican group, not without the support of Democratic elements, seemed on the verge of enacting legislation that would direct an investigation of the loyalty fitness of each employee. It was against this background that the personnel policies of the Truman Administration took shape. When Mr. Truman announced the Loyalty Order in 1947, it may be that he did so at least in part to deflate an issue that his political opponents had developed with a high degree of success, and perhaps with the intent of forestalling even yet more drastic action by Congress. Whether or not the Order had those purposes, they were in any event among its effects. The Administration's loyalty program, said Representative (now Senator) Mundt, "is almost precisely that which the House Committee on Un-American Activities has been advocating for at least four years"; and Chairman Reece of the Republican National Committee expressed gratification that "the President, however belatedly, has adopted this important part of the program supported by the Republican party and its candidates in the 1946 campaign." [26]

After the Republican electoral success in 1946, the President had appointed a Temporary Commission on Employee Loyalty, with instructions to study the matter and to recommend action. The Commission, composed of representatives of six federal agencies, duly reported that a loyalty program should be initiated. In reaching this conclusion, it indicated the following significant judgment:

> "While the Commission believes that the employment of disloyal or subversive persons presents more than a speculative threat to our system of government, it is unable, based on the facts presented to it, to state with any degree of certainty how far-reaching that threat is." [27]

Today the dimensions of the threat may be set forth with considerably larger confidence. As of June 30, 1949, inquiries into the loyalty of 2,541,717 federal employees or would-be employees had been made, and presumably additional hundreds of thousands were made in following months. In all but some twelve thousand of these perhaps 3,000,000 cases, the FBI reported that the records were so spotless that no derogatory information whatsoever appeared. By May 31, 1950, the loyalty boards had received reports of 11,844 cases in which investigation had disclosed unfavorable data of some sort—about four-tenths of one percent of the total cases checked by the investigating agency. Upon further full investigation, the great bulk of these cases were found to raise no serious question concerning loyalty. They were closed with favorable findings. In only 478 cases of incumbent employees and appointees was there enough doubt in the minds of the loyalty boards to warrant their making determinations of ineligibility. One hundred fifty-one of the adverse judgments of loyalty boards had already been reversed on appeal to the Loyalty Review Board, and 102 of the cases were still pending. If we assume that every single one of the as yet unreversed ineligible determinations were to be sustained on appeal and if we add to this number all the 1,068 employees who left the federal service after investigation of them had been completed but before adjudication of their cases, we have a gross figure of 1,395, with another 530 cases still unacted upon and resting on the boards' dockets. Since this includes persons who resigned or retired for reasons wholly unrelated to the loyalty program, as well as everyone who has been adjudged to be potentially disloyal, this total surely gives us a workable notion of the dimensions that the Temporary Commission on Employee Loyalty could not state. Indeed, after the investigation of well over 2,000,000 employees had been completed, then Attorney General Clark publicly acknowledged, "While highly paid investigators have used millions of dollars of the people's money, as yet they have failed to uncover one Communist presently working for the Federal Government." [23]

The fact is, of course, that the really dangerous culprits, if they do exist, are not likely to be found by the dragnet methods that are perforce the means of executing a wholesale program. It is no criticism of the investigating agency to say that its loyalty inquiries will lead in the main to persons who are merely rebelliously unconventional or outspokenly assertive or even obnoxiously opinionated. The mass investigation of federal personnel will rarely expose the furtive, the corrupt, and the conspiratorial.

In point of fact, not a single individual who has been dismissed under the loyalty program has been indicted or prosecuted for traitorous misconduct that the investigation brought to light. It is worth recalling that the flamboyantly publicized prosecutions of recent years did not grow out of loyalty proceedings. Alger Hiss was fully "cleared" by Secretary of State Stettinius and Secretary of State Byrnes after an investigation of essentially the same type as the present loyalty probes, except that his was perhaps more intensively conducted. Judith Coplon, a Department of Justice employee who has been found to have conspired with a foreign agent, was fully investigated by the FBI before she was assigned to the confidential duties of her job. The investigation at that time disclosed none of the behavior or ideological patterns that led ultimately to her involvement in a crime. Neither suspicion nor detection of her acts arose from the loyalty program, but rather from the internal operations of the Department of Justice itself.

Cases like these suggest that the approach of the present Order might well be supplanted by other more functional steps. One must commence by fully accepting the proposition that the Federal Government, even more than most employers, is entitled to demand and receive the loyalty of those who serve it. The question now presented relates only to the method the Government should use in assuring that its employees are in fact loyal. The present method, though it uncovers little evidence of disloyalty, leaves many wounds and produces much wreckage. The social consequences are too great to permit us to ignore the ineffectiveness of the program as it is now conceived.

The personal beliefs of the seismologist, the poultry disease specialist, and the oceanographer should cease being a matter of governmental concern except as they may be objectively reflected in their actions. Once again there needs to be an emphatic differentiation between the loyalty program and the security program discussed in an earlier chapter. The security program involves persons in whom we wish to have the fullest confidence because of the nature of their responsibilities. Confidence cannot coexist with any serious risk that national safety might be jeopardized by unauthorized behavior or speech. The risk may arise from a job incumbent's character or his personality or his associations; and if we perceive the risk, we may simply choose not to take it. In other words, when we withhold "security clearance" we make no finding that otherwise an undesired event will surely come to pass; we merely find that there is an undesirable *possibility* and we seek to avoid even the possibility, let alone the actuality. But the loyalty program is differently oriented. It deals neither with "sensitive agencies" nor with "sensitive jobs." It involves no findings that in some undefined future there may be an improper transmittal of "government secrets," because most of those who are affected by the Loyalty Order know of none. Their work brings them in contact with no matters of national defense or international politics. The Public Health Service student of syphilis and the attorney of the Home Owners Loan Corporation have at least one thing in common: neither one, by virtue of his work, knows anything that even the most vigorous Russo-phobe would fear to have him tell his friends and relatives. When we concern ourselves with the loyalty of that sort of federal employee, there-fore, we deal not with an issue of trustworthiness related to the nation's safety. We deal, instead, with a disloyalty that we must find to exist here and now, a present reality though un-related to present conduct, being evidenced only by opinions or "sympathetic associations." It is this focus that has caused the disillusioning difficulties of the loyalty program.

Those difficulties would be diminished if we ceased search-ing for "disloyalty" as a general abstraction and became con-cerned exclusively with "security." Concededly there are posi-tions outside the "sensitive agencies" that directly involve national safety. Occasionally an entire section or division of an organization may have occasion to deal with classified mat-ters or may be so immediately involved in the formulation of international policy as to render it "sensitive" even though the agency as a whole may not be so. Conceivably, for example, the Division of Territories and Island Possessions of the In-terior Department may be intimately connected with the preparation and execution of defense plans, including the lo-cation of military installations in outlying portions of the American commonwealth. If that be true, some or all of the personnel of that division may fall within the area of concern about "security." This does not mean, however, that instantly the same concern arises about all the remaining 30,000 em-ployees of the Interior Department, scattered among the Indian Arts and Crafts Board, the National Park Service, the Bureau of Land Management, the Fish and Wildlife Service, and all the other agencies that are segments of that depart-ment. Somewhat similarly, the Office of International Trade of the Commerce Department may conceivably have enough power over the flow of strategic supplies and over the conduct of economic warfare to warrant inquiry into the "security" of its personnel. But that is not likely to be true as to the Bu-reau of the Census or the Inland Waterways Corporation.

The solution here is to authorize the head of each department and agency to designate the units or particular positions in his department which he believes to be "sensitive." Persons who may be employed in these sensitive posts may properly be investigated in order that there may be full confidence in them. But as for the rest—the typists in the Veterans Administration or the Federal Housing Administration, the scientists in the Allergen Research Division or the Mycology and Disease Survey of the Bureau of Plant Industry—experience under the Loyalty Order demonstrates that constant peering over their shoulders endangers liberty without enhancing loyalty.

This is the administrative device that has been tried with reasonable success in Great Britain.[29] There the power is lodged in each Minister to decide what parts of his ministry require the equivalent of our security clearance. In all, about 100,000 jobs were identified as having security significance. The Admiralty, as has our Department of the Army, concluded that everyone, from the highest to the lowest, must be cleared. Other ministries found no "sensitive" jobs at all. And this is as it should be, for in the variety of modern governmental activities there is room for both extremes.

If this approach be adopted, it will not mean an abandonment of interest in the probity of "nonsensitive" personnel. It will mean merely that observations will be related to behavior rather than belief. Government employees who improperly discharge their duties, whether motivated by disloyalty or mere slovenliness of habit, should of course be identified and appropriately disciplined. This, however, is a matter of administration rather than of detection. The supervisory officials of a functioning unit can more readily determine a staff member's misconduct or carelessness than can even the most vigilant agent of the FBI. The responsibility for efficiency should rest squarely on them. They cannot fulfill their responsibility if they tolerate on their staffs employees who are not actively loyal to their jobs. As for misdeeds unrelated to the direct performance of an employee's work, reliance must be placed upon the excellent counterespionage staffs of federal investigating agencies. The thorough work of the Federal Bureau of Investigation has given that bureau the place of public esteem that it occupies. The inherent absurdities of the loyalty program threaten the FBI's deservedly high reputation, for its "loyalty probers" must expend their energies in recording the often ambiguous pettinesses of political expression rather than in uncovering criminality. Releasing the FBI from the thankless and fruitless work to which it is now assigned will enhance the nation's safety. The more broadly we define the limits of our concern with personnel security, the more thinly we must spread attention to it. As has been true so often in matters of public administration, the scattershot of the blunderbuss is less effective than the aimed bullet of the rifle.

More than one hundred and fifty years ago a great friend of American democracy, Edmund Burke, argued that while restraint upon liberty may sometimes be required if liberty itself is to survive, "it ought to be the constant aim of every wise public council to find out by cautious experiments, and cool rational endeavors, with how little, not how much, of this restraint the community can subsist; for liberty is a good to be improved, and not an evil to be lessened." Burke's words are as true today as when he uttered them in 1777. The country will be the stronger for discovering that the restraints of the present loyalty program exceed the needs of national preservation.

VII

The Universities and Security
Searches

THE area of personnel security proceedings has broadened, as has been seen in preceding chapters, until it now reaches scientists who themselves have no direct access to secret data. It is clear, too, that the Federal Government energetically concerns itself with personal or imputed beliefs as distinct from the observable behavior of its employees, in the hope that thus it will be assured of their loyalty. There is a relationship between these matters and the academic cloisters in which novitiate scientists are being educated for the tasks of the future.

Universities have traditionally been the chief centers of pure research in this country. Today Government laboratories and even industrial research laboratories are heavy contributors to fundamental knowledge, but it remains true that academic researchers tend to concentrate upon discovering the basic data, which others may then develop and apply in "practical" ways. Developmental research, unless it is to be devoted merely to elaborating the gadgetry of contemporary civilization, must draw upon the ever-growing stock pile of suggestive ideas and fresh facts which can most readily be supplied by investigators unconcerned with immediate results. Scientific applications may well be likened to superstructures which must rest upon the solid foundation of preceding experimentation.

During the war basic studies were subordinated to the pressing needs of the moment. The nation used its existing scientific resources to great advantage. Old ideas were exploited in new ways, and new techniques were devised to increase efficiency or productivity. But the times were too hectic to permit the questing, probing, restless experimentation of the fundamental scientist. Recognizing that past successes in applied science were not guarantors of future advance, the Federal Government in postwar years has markedly increased its support of university research. The annual research budget of the physics department of one eastern college has, for example, risen from a prewar $20,000 to a postwar $800,000, much of the increment coming from public funds. Perhaps at no previous time in American history have federal monies played so large a part in so many institutions as they do today.

That is not to say that governmental assistance is carelessly extended. In fact, it is given only on the basis of contracts, and with reference to specific projects. The two chief sources of funds have for some time been the Office of Naval Research and the AEC.[1] Before either one lets a contract, it carefully reviews a proposed research project in order to estimate its scientific soundness, the qualifications of the scientists who will participate in it, its relationship to the general areas in which federal support can be justified, and its likely contribution not only to knowledge but to the nation's pool of scientific manpower.[2] The Office of Naval Research—ONR— sponsors well over a thousand separate nonsecret projects, scattered among fully two hundred institutions and ranging in subject matter from astronomy to viruses. Students of the army ant and of white dwarf stars are alike aided by ONR funds, as are many others whose researches will not necessarily (though they may conceivably) have eventual naval applications. ONR's present investment in basic research is larger than the entire prewar national expenditure for that purpose, and accounts now for perhaps as much as 40 per cent of what America spends on "pure science." With similar breadth of interest, the Atomic Energy Commission has entered into contracts with nearly a hundred separate educational institutions, calling for "unclassified" research in chemistry, mathematics, metallurgy, physics, biology, and medicine. Projects involve such diverse topics as research on corrosion of alloys, the effects of the irradiation of peanut seed, and the characteristics and physiological consequence of "flash burns" resulting from bomb explosions. Sometimes the AEC and the ONR join forces to support enlightening but only remotely military studies of a physiological as well as of a physical character.[3]

In addition to these contracts that involve no element of secrecy whatsoever, both the AEC and the ONR occasionally request a university staff to assume responsibility for a "classified" project, or, as one outstanding administrator of academic science put it, "do a little favor here and there." Each "favor" brings into the university the same security apparatus that is operative in laboratories like those at Los Alamos or Oak Ridge. Access to a part of the laboratory and its equipment must be barred. Only those who are "cleared" may work in the project. The problems encountered during the investigation may not be discussed with colleagues. The results of the researches may not be freely communicated either to students or to faculty members.

Taking note of these consequences, the AEC's director of research, Kenneth S. Pitzer, himself a former professor of chemistry, has expressed the hope that secret atomic energy research can be kept out of university laboratories; even the small projects, he has asserted, build up walls that destroy the open freedom of academic intercourse.[4] It may be well here, by way of underscoring Dr. Pitzer's concern, to repeat that published papers are only one (and not always the most important) means of scientific communication. "To an extent much larger than is realized," writes a prominent educator and researcher, "the transference of scientific ideas from one set of scientific workers to another is effected by means of visits, personal contacts, and letters . . . Almost every visit of a scientist from one laboratory to his colleagues in another results in the introduction of a new piece of information or point of view that no amount of reading had managed to effect." [5] When a part of the university is, so to speak, blocked off from the rest, the university to that extent ceases generating the ideas and spreading the learning for which it exists.

Some of the leading institutions with large endowments are so fearful lest secrecy invade their halls that they forbid acceptance of any "classified" work whatsoever. But this policy of excluding secret work is both expensive and difficult to maintain. In our era research equipment and staffs cost dearly, and private benefactors are a fast-vanishing tribe. Universities that desire to expand their facilities and their personnel therefore eagerly snatch at subsidies in the form of research contracts. Once the subsidized expansion has occurred with consequent changes in the institution's financial structure and economy, the university must perforce become highly sensitive to anything that threatens continuation of its enlarged scientific program. A few, like the University of Chicago, guard against the possibility of later dislocations by limiting the portion of research which may be governmentally financed. In those schools a withdrawal of support or an irreconcilable clash of philosophy will not destroy the university's scientific experiments altogether, but will simply lead to jettisoning the projects with public funds. Others may face the collapse of their entire program unless they accept whatever conditions may be attached to the grants. During the fiscal year 1949 federal expenditures for research of all kinds in educational institutions exceeded $200,000,000.[6] So staggeringly large a sum constitutes a major share of the income of American colleges and creates a condition of dependency which may be unwholesome.

It ought to be said forthrightly that the Office of Naval Research and the Atomic Energy Commission have been exemplary in their behavior toward universities. The terms of the contracts themselves may sometimes narrow the field of research in ways that impair its educational value. But neither one of these major contracting authorities has as yet sought to fasten control upon institutional policies by manipulating the purse strings, and neither one has indicated a desire to do so in the future.

Atmospheres and personnel may, however, change; and with the changes may come new attitudes. Indeed, even without formal declarations, novel trends of educational policy may already be discerned. Consider as a symptomatic instance the case of Cornell University.

When World War II ended, Cornell established a firm policy of prohibiting classified research projects on the campus, as distinguished from the Cornell Aeronautical Laboratory in Buffalo, where classified projects were freely accepted but were obviously remote from the academic activities of the university. This policy, as summarized by former President and Chancellor Edmund E. Day, rested on the following reasoning:

"Since the primary functions of a university are the acquisition and dissemination of knowledge, university research should be such that the results may be freely published. Incidentally, this is a favorable condition for the efficient conduct of fundamental research inasmuch as significant progress is generally the result of the interplay of several minds. A

179

university campus is particularly suited to such interchange of ideas between those who have specialized in related fields. There is, then, the possibility that when university research is classified the value of the results emanating from it may be reduced because of the impediment to free exchange of discussion between those directly engaged and other scientists of our University staff and, as well, between investigators on our own campus and others of corresponding interests on the campuses of other universities. The difficulties of obtaining security clearance and of the physical handling and filing of papers were also cited. Then, too, there is the possibility that certain of our scientists who have had extensive experience with the effects of classification of research during the war will be highly resistant to working under such conditions in peacetime. One further point is the extreme desirability of retaining publication rights for research studies made by our staff which would be one of the first items eliminated in the case of classified projects, except after obtaining approval of the sponsoring agency." [7]

But on September 9, 1948, Dr. Day, having noted "the pressures we are receiving from government agencies to modify our policy to some extent," announced a relaxation of the prohibition. In connection with several existing projects that had commenced on a nonsecret basis, he said, "the government agency concerned has expressed the view that either immediately or within a short period of time, the work under the project must become classified. We may anticipate, I believe, that such pressures will be increased in the case of other projects as time goes on." After reaffirming the basic policy of abstaining from acceptance of classified projects on the campus "under true peacetime conditions," Dr. Day stated the conclusion that "current international conditions (the cold war) and the general national defense and security 'atmosphere' of the country at the present time justify some relaxation in strict adherence to that policy, an action bringing us into general conformity with the policy already existent in many other universities." By way of protecting individuals against unwelcome intrusion of "security clearance," the flat stipulation was added that "no professor or graduate student will in any way be *forced* to associate himself with such [classified] work, whether it shall be a project just inaugurated or one involving an extension of previous work."

Cornell is not among the more impoverished American universities. With its large scientific faculties and staffs, it can well afford to tolerate a few individuals who choose not to work under secrecy restrictions or who might be barred from doing so because they could not obtain the necessary clearance. Whether all educational institutions will feel able to exhibit equal tolerance is highly uncertain. A smaller school, with only a few men in each of its departments and with heavy dependence upon government research funds, may be under irresistible compulsion to rid itself of an individual who is not "adaptable" and who "does not fit into the school's research program." The principles of academic freedom, from which all great universities derive their strength, are not protected with invariable vigor by administrators concerned with urgent financial problems.

There have been occasional and untypical instances in which university scientists have unprotestingly accepted the denial of clearance to junior colleagues, though privately expressing the opinion that the denials were unfair; and when the young men were subsequently dropped from their posts, no outcry has been heard despite acknowledgment that the affected individuals were well qualified as scientists. It is true that as yet few difficulties have been encountered by persons who have "tenure" in their jobs. Those (and in absolute numbers they have not been many) whose academic status has been impaired by clearance problems have been the beginners—

the instructors and staff assistants whose budding careers have not yet led them to professorial rank. One cannot assert that the inroads upon academic communities have been serious. It is well to remember, however, that invasions of freedom usually have inconspicuous beginnings. The danger lies in the precedent which those inconspicuous beginnings sometimes serve to establish. Once the fire begins to burn, it stubbornly resists being extinguished. We are still far from emulating the passion for academic destruction that swept over Germany in the thirties. It is nevertheless worth recalling that the intrusion of academically irrelevant considerations into German universities commenced with the unnoticed and unprotested removal of instructors, coupled with reassurances to professors who had tenure.[8] In the short space of the next four years, however, there were "retirements" of three times as many science professors as had dropped from the ranks in the previous four years.[9]

In the main the records of American universities in a time of tension have been excellent. Occasionally, as at one of the most distinguished of the eastern universities, a department head is heard to remark that he will not engage, even for wholly unclassified research or teaching, any man who has been denied clearance at a Government laboratory—and this even though, as earlier chapters have sought to make clear, a denial of clearance does not necessarily involve a finding of reprehensibility. This readiness to outstrip the necessities of the situation does not find general support among university scientists. Most of them, on the contrary, agree with President Conant of Harvard in saying: "The government, of course, must see to it that those who are employed in positions of responsibility and trust are persons of intelligence, discretion and unswerving loyalty to the national interest. But in disqualifying others we should proceed with the greatest caution . . . The criteria for joining a community of scholars are in some ways unique. They are not to be confused with the requirements of a federal bureau. For example, I can imagine a naive scientist or a philosopher with strong loyalties to the advancement of civilization and the unity of the world who would be a questionable asset to a government department charged with negotiations with other nations; the same man, on the other hand, because of his professional competence might be extremely valuable to a university." [10]

Academic scientists are not confident, however, that President Conant's views will everywhere prevail. It has already been suggested that the boldly creative scientist may be equally bold, though perhaps not equally creative or informed, about social problems. Men of this stamp are the ones most likely to encounter the doubtings that have previously been described. They may be forgiven for fearing, today, that the repercussions of clearance difficulties will be felt even in the isolated groves of educational institutions. That, at least in part, is why some of the university men decline to participate in either classified or unclassified projects supported by AEC or ONR funds; they are still to be convinced that the security check and the loyalty test will not be applied indiscriminately. And if one asks why reputable people should be concerned, there are only two answers to be given. The first is that many an honorable man cherishes his privacy and will not willingly see it invaded. The other is that the margin of error in our checks and tests is still so great that an entirely innocent person may prefer, if he has an effective choice, to avoid them altogether.

Once again, our major concern is not one of sentiment. In reaching conclusions about the extent to which personnel security and similar procedures may safely be intruded into institutions devoted to learning and teaching, we may properly be guided by an enlightened self-interest. In a sense, and indeed in a very vital sense, the happiness and well-being of every individual in society is a concern of all; and a truly far-sighted people must weigh against any general policy the personalized distress it may cause even a small number. Here, however, the question of societal advantage may be considered in less immediately human terms. The issue may be reduced to this: Are broad scientific gains probable if the availability of academic research workers for particular problems is conditioned not solely upon their having unassailably sound professional qualifications, but also upon their having unassailably "correct" political attitudes?

One of the dangers which, though not as yet exhibited, inevitably inheres in the military's large-dimensional domination of scientific research in universities is that the idiosyncratic and the unpredictable may come to be deprecated, perhaps in the end ruled out altogether. Opposition to technological change has, historically, been a characteristic of the profession of arms in every country. Despite the evidence to the contrary which is today being provided by the Office of Naval Research, it has as a rule been the pressure of civilians on the military, and not the reverse, which has led to encouraging scientific experimentation and adopting its fruits.[11] Even if this were not so, the conventional rigidities of military supervision looking toward "efficiency," coupled with distrust of the innovator and the heretic, could ultimately standardize scientific effort rather than energize it. General Sir Ian Hamilton, himself a military administrator of stature in the preceding generation, has well said: "In precise proportion as highly organized systems increase the cohesion and momentum of their mass, so they must flatten out the idiosyncrasies and clog the alertness of each of the component particles of that mass. In precise proportion as the machine becomes effective, so do the chances of evolving an engineer of initiative become smaller." [12]

This is not an argument for chaos. It is an argument, rather, for watchfulness lest practices that are functionally appropriate in only a limited set of circumstances and in only a specialized type of organization creep imperceptibly and needlessly into the world of scholarship. Freedom of the individual scientific worker to choose his own subject may result in a somewhat haphazard pattern of activity, but, if we may believe so great an authority as Enrico Fermi, it is "the only way to insure that no important line of attack is neglected." [13] For that reason, amongst others, governmental support of academic research, whether administered as now through the Navy and the Atomic Energy Commission or, as will soon be the case, through the newly created National Science Foundation,[14] must not be permitted to compromise the independence of university staffs. The daring that leads men into the realms of the unknown cannot be regimented. The mental qualities involved in envisioning and planning projects that may add to human knowledge cannot be prescribed. The capacity to master the techniques of scientific research cannot be confused with the capacity to think respectably, or not at all, about social problems.

To "get research done" requires much more than merely setting aside a given sum of money. Able men and women can be aided by the facilities the money can provide; the money alone produces no results whatsoever. Dr. Alan Gregg, distinguished director of medical sciences at the Rockefeller Foundation and chairman of the AEC's Advisory Committee on Biology and Medicine, has warned that "unless young men can plan lives as research men, they won't go into it or stay in it. Fellowships for a year or so are not enough inducement. One of the difficulties of the machine age is that men fall to treating each other like machines. Good scientific work can be done by our already experienced investigators when money is provided for instruments, consumable supplies, technical assistants, etc. Trained and able investigators are products of good education, of long training in the atmosphere of scholarly devotion to research and of the conviction that their lives can be spent decently in such careers." [15]

Today, as for some years past, manpower has been the limiting resource in the nation's research endeavors. In point of numbers the United States lacks the bachelors and doctors of science to man the projects that await attention.[16] In terms of top-notch ability the discrepancy between demand and supply is even greater. In order to close that gap, the Federal Government has of late years provided research fellowships for persons whose training and background support the belief that further educational experience will be in the public interest. The Public Health Service, for example, grants fellowships for advanced study in medicine and related subjects. Similarly, and in order to help build up a pool of men trained to participate in physical, biological, or medical research, the Atomic Energy Commission began an extensive fellowship program.[17] The one-year AEC grants range from $1,600, for an unmarried person who has not yet earned a graduate degree, to a top of $4,000, for a married man with two or more dependents who holds a Ph.D. or M.D. degree. During the academic year 1949–1950, 421 young men and women, selected after comparative evaluation of their records and potentialities, were engaged in studies, most of which were wholly nonsecret, on such matters as the effect of radiation on viruses, the biophysics of the nervous system, and the fundamental physical aspects of structure. Despite the range of the projects, all in some way related to the AEC's concerns.[18] A study of the auditory mechanism, for example, has been held to have too incidental a relationship to the atomic energy program to warrant AEC support, though there were no doubts about the value of the project or the capability of the applicant. On the other hand, a fellowship has been granted for work on the theory and design of high-speed calculators, a matter that may have importance for mathematical work in general as well as for AEC researchers in particular.

Whatever doubts American universities entertain about the desirability of Government support of academic research have been sharply accentuated by recent events in the AEC's fellowship program.

When the program was launched, the AEC, availing itself of the statutory power to utilize advisory bodies, turned over to the National Research Council the task of selecting the fellows to whom grants should be made. The National Research Council, or NRC, is an adjunct of the National Academy of Sciences. Since the First World War it has functioned as a scientific adviser to the nation upon the request of the Government. For some years it has administered fellowship programs for the Rockefeller Foundation, the American Cancer Society, the National Foundation for Infantile Paralysis, and the like. In doing so it has developed to a high state of perfection the organization of "fellowship boards" composed of eminent educators.[19] These boards have passed on the qualifications of each applicant, the utility of his proposed studies, and the capacity of any particular institution to furnish postgraduate training of the sort he sought. In considering an application, the fellowship boards have examined confidential reports on the candidate from experienced scientists who were familiar with him. In some instances these reports were supplemented by personal interviews. The fellowship board's decision, if favorable, reflected its judgment that the candidate was capable of making substantial contributions to scientific progress and that a grant of a fellowship would therefore be in the public interest. That the Council's methods are successful is attested by the present eminence of some of those whom it has selected for fellowships in the past, including E. O. Lawrence, the inventor of the cyclotron, Samuel K. Allison, director of the Institute for Nuclear Studies at the University of Chicago, W. V. Houston, president of Rice Institute, and Norris E. Bradbury, director of the Los Alamos laboratory.

From the outset of the program it has of course been hoped that, in one way or another, scientists trained through the fellowship grants might ultimately be useful in the nation's program of atomic energy research. But no commitment as to future employment has been made either by the AEC, any of its contractors, or the fellowship holder himself. Many of the fellows will no doubt continue in research outside the AEC's scope of concern, and some of those who may some day work directly on AEC matters will probably be engaged in entirely unclassified experiments.

Occasionally in the past a project has involved access to secret material, or even working in an AEC installation.[20] Of the 421 fellows in 1949–1950, only 30 were engaged in research that involved restricted data. In instances of that sort, the fellow has had to obtain the usual AEC security clearance, after the intensive FBI investigation and the various analytical procedures that are set in motion in connection with any full-time employee in a restricted area. On the other hand, soon after launching the fellowship program the Commission had established the policy of not requiring clearance where there was no element of access to restricted data or areas. Among the reasons it had advanced for this policy were that "we will obtain more qualified fellows and achieve fuller cooperation from the scientific community of this country than would be the case if we adopted the principle of requiring security clearances at a time when it is contemplated that fellows will not have access to restricted data." Moreover, the Commission added, "it must be recognized that security investigations are costly, and that the cost of these investigations will be kept to a minimum when they are carried out only when the particular person is to have access to restricted data. It is probable that many of the fellows will always be engaged in unclassified work, so that the costs of security investigations as to them could well be an unnecessary expense to the Government if undertaken prior to the award of the fellowships." [21]

So matters stood until May 1949, when a radio sensationalist disclosed that a young Communist had been granted a $1,600 fellowship to do work at the University of North Carolina. The young man was to study for a doctorate in the field of general relativity, a project without military or commercial applications of any sort and wholly without access to restricted materials; but these were mere details that were lost among the exclamation marks. Almost simultaneously it was learned that a postdoctoral fellow at Harvard, working on an entirely nonsecret endocrinological study, had once attended some Communist meetings, though he subsequently denied vigorously and apparently convincingly that he was in fact a member or supporter of that group.

Then the storm broke. The revelation that the Atomic Energy Commission was supporting the training of suspected Communists created a furious demand that remedial steps be taken. The juxtaposition of "atom" and "Communist" stimulated a fervor of response which was undiminished by the sobering facts that the fellowships involved no danger to secrets of any description. In vain did Alfred Newton Richards, vice-president of the University of Pennsylvania and president of the National Academy of Sciences, urge that educating an exceptionally qualified person, even if a Communist, "will have added one more to the group—now far too small —of those capable of utilizing knowledge of nuclear energy and of its products in the advancement of medicine, biology, agriculture, and, at need, could release for Government classified service another who possessed no disqualifications. The country will have been the gainer by his training." [22] In vain was this thought echoed by Detlev W. Bronk, president of the

Johns Hopkins University and chairman of the National Research Council. In vain did J. Robert Oppenheimer hammer at the proposition that many discoveries in the past, basic to the present work of the AEC, had been made by persons who could not be cleared, and great discoveries in the future might also come from men whose political purity might be challenged.[23] In vain did President James B. Conant of Harvard object that if all fellows were subjected to standard clearance requirements, the "vast amount of checking and personal investigation" would soon create an "atmosphere of distrust and suspicion in the scientific world" far outweighing any possible gain.[24] In vain did President Lee A. DuBridge of California Institute of Technology assert that "to extend political investigations to young students working in non-secret fields where there is no question of national security involved at all I think is contrary to American principles of democracy"; trying to sift out communistically inclined applicants "would bring the basic ideas of a police state into American youth," and would entail the use of methods "far more dangerous than the small risk of having an occasional Communist on the fellowship rolls." [25] In vain did the executive committee of the American Institute of Physics protest that investigating AEC fellows as though they were AEC employees "would be an unnecessary extension to the field of education of measures appropriate only in secret work." [26]

When all the warnings had been sounded, they were simply ignored. Congress proceeded to enact into law the proposition that no AEC fellowship funds shall be given to "any person who advocates or who is a member of an organization or party that advocates the overthrow of the Government of the United States by force or violence *or with respect to whom the Commission finds, upon investigation and report by the Federal Bureau of Investigation on the character, associations, and loyalty of whom, that reasonable grounds exist for belief that such person is disloyal to the Government of the United States."* [27] Thus the measure as adopted went beyond the mere barring of Communists from the fellowship rolls, but extended to all applicants for this type of grant the same sort of FBI inquiry and agency determination made in loyalty cases involving regular federal employees. The chief difference is that a federal employee or would-be employee is accorded the privilege of a hearing, albeit an imperfect one, before the dire finding of disloyalty is made. The youthful seeker of scientific training, on the other hand, may suffer rejection and its long-lasting consequences without ever having opportunity to interpose a defense.

What were the reasons for so drastic an action in dealing with a problem of so little real substance? No one seriously supposes that there is a significant Communist element among the applicants for AEC fellowships. The applicants are men and women who have already achieved a measure of academic distinction and who carry the endorsements of experienced scientists and teachers. Few American university professors have discerned any great inroads of communism upon today's student bodies, and it seems particularly unlikely that "infiltration" is considerable among the young people who have devoted themselves to intensive scholarship rather than to the excitements of contemporary politics. That group, as described by Dr. Richards, "is made up of unusual individuals. Their mental qualifications have been found to be exceptionally high; commendatory references have been obtained from their professors with whom they have worked; they have become enamored of science and are preparing to devote their lives to it." [28] Before the AEC had really made up its mind whether or not applicants for fellowships should be investigated, and long before the Congress had turned its attention to the problem, 151 seekers of fellowships had in fact been fully investigated by the FBI. One investigation turned up the open and avowed North Carolina Communist whose $1,600 grant touched off the fireworks; another apparently established the "Communist affiliation" which the Harvard fellow seems fairly well to have disproved when given a chance; and two more suggested some sort of "Communist association" short of affiliation.[29] If this same relationship is maintained in the future, it appears that full field investigations of applicants for grants in nonsecret fields will produce one sort or another of "derogatory information" in about 2.5 per cent of all these cases; of this derogatory information, only half will be of a serious nature; and only half of the seriously derogatory information will continue to seem serious if subjected to the test of a hearing; so that when all is done, only two-thirds of one per cent of the applicants will remain under a heavy cloud of doubt concerning loyalty.[30] What impulses moved Congress to unlimber such heavy artillery to blast so minor a target?

The impulses were of course diverse. One position that was stated and restated was that public funds should not be spent to educate a Communist, who by definition is deemed a conspirator against the Government. This position overlooks the fact that the barrier which Congress erected will keep out not only Communists but also those who may be "disloyal" in the much broadened sense. In any event, the question here was clearly not one of economy. The loyalty checks that Congress has commanded will at the most conservative estimate cost annually no less than $50,000 in direct expenses of investigation, plus the time and attention of security staffs which have important duties elsewhere. The basic issue was not whether money should be spent. The question was, simply, whether a politically objectionable person should be permitted within the area of the expenditure. Senator Hickenlooper explained the matter succinctly when he said: "I think you can say it in a nutshell: I do not believe the American people will stand for the education of a Communist with public money." [31] Congressman Durham, vice chairman of the Joint Congressional Committee on Atomic Energy, wished "to keep education as free as possible," but the grant of a fellowship to a Communist caused him to exclaim: "The country is just not going with us and we have to go to the people and tax them to get appropriations." [32] Senator Millikin emphasized again that the American people "have the notion, for which considerable support can be developed, that the United States should not be spending the taxpayers' money to educate anyone who joins a conspiracy against the United States." [33]

This view was put to the test when the North Carolina Communist disclosed that he had completed his education with the support of the benefits received under the "G.I. bill." When President DuBridge observed that nobody had complained at that time, or even now, about the fact that a Government educational subsidy had been paid to a Communist, Senator Hickenlooper distinguished the cases by saying: "The G.I. educational bill is based on the theory of an earned stipend. It is the payment for something that has been earned prior to that time." Dr. DuBridge made the immediate rejoinder that so, too, in a sense, a national research fellowship is an earned stipend. The fellow undertakes to develop his skill and his brains for the nation's benefit; "he has earned his education by his ability, as proven in his previous work, and he is doing a service to the country by training himself." [34]

It was Senator Hickenlooper, not Dr. DuBridge, whose views prevailed. Dr. DuBridge disliked the idea that "we are doing a favor to these fellowship candidates by giving them a fellowship." He preferred to think that "the country is getting a good bargain in spending money to train these men who will be important in the future leadership of science." The Congress of the United States thought otherwise.

Here is a clear-cut issue. Those whose profession involves

a certain attentiveness to public opinion gave one answer. Those who did not have to face the electorate concluded that the contrary answer was the only sound one. "The people and the Government of the United States have a stake in scientific discovery and invention," said Dr. Oppenheimer, "and it is for this stake, rather than as an act of benevolence toward the recipients of the grants-in-aid, that one must look for justification for having a fellowship program at all." President Conant was sure that no great harm would result even if a Communist did become a fellow, for if he ever sought access to confidential information, he would have to be investigated fully. Meanwhile, "if such a man continues in the field of pure science he may make important contributions." Dr. Gregg stated what he regarded as an axiom, "that this condescension on the part of the Government to give these young men an opportunity is seriously inaccurate and almost to the point of being quite a false view of the situation. We are looking for brains and we are looking for character and when we can find them, it is as good as a business deal with both sides profiting . . . Now, I would not care to open a fellowship program under circumstances that would dissuade a seriously large number of applicants from applying. I would not open with a note of distrust for the simple reason that young men who have their careers to make are pretty concerned about it and if they suspect something that they do not like and can go elsewhere, and thereby avoid it, you will not have them nibbling at it and you will not have a chance to get them." [35]

This last remark suggests another one of the major divisions of opinion between the members of Congress and the members of academic or scientific communities. Throughout the hearings the former made clear their opinion that no true American would be repelled by a requirement of oaths and subsequent official investigations into his character, opinions, and associations. Only conspiratorial enemies of the nation, and perhaps a few others who were willfully perverse, would hesitate to subject themselves to scrutinizing of their "loyalty." A different position was taken, as with a single voice, by those whose work had brought them in closer touch with young intellectuals than with practical politics.[36] All of them in one form or another stressed the experimentalism of youthful minds and the likelihood that the unorthodoxy of youth would be modified by later experience. All of them felt that many able men would choose not to place their careers in jeopardy by risking the unpredictabilities of a loyalty test. All of them feared that the very process of investigation, involving the questioning of schoolmates and teachers and neighbors, would engender suspicions and uncertainties that would have a seriously adverse effect "on both the atmosphere of our educational institutions and the outlook of one age group of the entire nation." [37]

No one can say with utter assurance which of the conflicting positions is correct. Many professors, however, have been told by able students that they shun federal service today because a careless rejection of them would produce lasting damage to their professional standing. In all probability the fear of rejection is rarely well founded. But men are moved to act (or refrain from acting) not only by reality but also by their images of reality. In a considerable number of instances young men's images of reality have caused highly trained and thoroughly qualified social scientists to withdraw themselves from the potential supply of governmental personnel. The scientific fellowship program will almost certainly suffer from the same sort of slow but debilitating bleeding.

Here it is perhaps well to note yet another division between the Congressmen on the one hand and most of the scientists and educators on the other. The Congressmen tended to doubt that the progress of science would be retarded by excluding the politically detested. They believed, in sum, that there would be no "bleeding" of the fellowship program if Communists and their supposed followers were kept out of it.

Part of this belief reflects the almost universal sentiment that one who does not share our own particular convictions must be a fool or a knave, or perhaps both. A comment of the ordinarily temperate Senator McMahon is illustrative. In discussing the young North Carolina Communist, who had publicly declared his disbelief that the Communist Party (United States) is controlled from abroad, Senator McMahon remarked: "He says he is in the pursuit of truth. And what more palatable and obvious factor is there, Doctor, to you and to me, than that the Communist Party in this country is part of an international Communist Party and an international conspiracy? . . . So if this fellow is so dense as not to see that, he must be a boob, and he is not worth anything . . . I do not mean that a great scientist has to be a conformist in his political views, and must think exactly as I think. I certainly do not mean that . . . [But] this statement about communism not being a national conspiracy, seems to me to be such a statement as to indicate that he is not very bright,"—and therefore should not have been granted a fellowship.[38] It is only fair to the Senator to add that his expression was enthusiastically seconded by Dr. Bronk, a distinguished scientist himself and the head of a major university.

The danger in accepting the view just quoted is that, despite disclaimers of insistence upon conformity, we may tend to decide whether a man is a "boob" or "not very bright" as a scientist by examining his opinions in nonscientific areas. The conclusion of Senator McMahon and Dr. Bronk that the Communist Party U.S.A. is a segment of an international combination seems to me to be unassailably based on the available evidence. But there must be many other propositions which Senator McMahon, Dr. Bronk, and I accept as palpably correct and which might nevertheless be contradicted by other mentally competent and disinterested persons. There is peril in insisting that anyone who rejects our own perceptions of truth in matters about which we deeply care must stand condemned. One need only instance Joliot-Curie, the French Communist nuclear physicist who discovered radioactive isotopes, or Lodge, the great British physicist whose faith in spiritualistic phenomena has been shared by few serious thinkers, or Eddington, whose religioscientific ruminations have not commanded as much respect as have his astronomical studies—one need only instance such men to realize that a scientist, like most of the rest of us, can be highly qualified in his own work and yet by some be thought a "boob" when away from it.

Finally, in connection with the fellowship program, one must make especial note of the layman's inability to distinguish between secret and nonsecret scientific work. Many people today equate the nation's strength with its ability to perform scientific miracles. For most citizens, including most members of the Congress, all science is a mystery. The beginning and the ending of terra incognita are but dimly understood, and the methods of exploration are little known. How else can one explain some of the concern lest a Communist-minded youth receive aid in tumor research or in studying plant nutrition? Surely no one presumes that a belief in free enterprise is a necessary qualification for intelligent investigation of the effects of irradiation on animal tissue; and it seems unlikely that new discoveries about cancer will be declared unsuitable for use in this country unless the discoverer can gain a security clearance. But preoccupation with the relatively small area of secrecy in science seems to stimulate an unreasoning fidgetiness about all scientific endeavor. Senator Knowland, for example, discerned what he described as "the calculated risk" that one of the AEC fellows working in a nonsecret research field might not only learn "some important scientific fact in medicine, or something else," but might also "hit upon a 'superduper' atom bomb, and be off to Russia, as Mr. Eisler was, on a boat, trying to get out of the jurisdiction of this country. And from the calculated risk point of view, he might be just the missing link to furnish information to an international conspiracy which has as its avowed purpose the destruction of the Republic and all that it represents." [39] Similarly, his colleague Senator Millikin was sharp in his reaction to the opinion that an unnecessary expense was involved in investigating the AEC fellows who would be working in nonsecret projects. "I would rather spend a hundred thousand dollars," said the Senator, "or several times a hundred thousand dollars to keep any conspirator against the United States Government out of the field of atomic energy. Put your own dollar sign on it. Write your own check on that." [40]

The immediate consequences of loyalty tests for AEC fellows can be quickly though not happily described. The National Academy of Sciences and the National Research Council on November 2, 1949, notified the Atomic Energy Commission that they no longer desired to accept responsibility for the altered fellowship program. The requirement of FBI investigations of those who neither work on secret material nor are directly preparing for work on AEC projects was regarded as "ill-advised." It raised "grave doubts whether the continuance of the Atomic Energy Commission fellowship program thus restricted is in the national interest." [41] This communication launched a series of further discussions. The AEC was unwilling, as a Government agency, to administer its own fellowship program, feeling quite properly that a scientific or educational organization should be in charge of the matter; it recognized, moreover, that it could hope for little success in effectuating a program that the scientific community would not fully support. Finally, the National Academy of Sciences was prevailed upon to authorize the National

Research Council to administer a drastically reduced AEC fellowship program, but only for a single year. For the academic year 1950–1951 the NRC would recommend no new predoctoral fellowships. Postdoctoral fellowships became available only for advanced training in fields of secret work or in problems that require access to restricted data. The fields of study were limited to those intimately related to the AEC program, such as the chemistry of the elements in the fission-products range. No medical, biological, or biophysical studies were to be undertaken unless they required the use of the special facilities available in the AEC installations or involved access to restricted data; the range of projects was thus narrowed to such matters as the development of radiation instruments as applied to biological and health physics problems of a classified nature.[42] Subsequently the AEC launched a greatly reduced predoctoral fellowship program for less advanced research in the biological and physical sciences. Administration was organized on a regional basis. The distinguishing feature of the new predoctoral program is that "the subjects of research must be sufficiently closely related to atomic energy to justify a presumption that the candidate, upon completion of his studies, will be especially suited for employment by the AEC or one of its contractors."[43]

The constricted fellowship programs led to making perhaps 75 new postdoctoral awards and 140 predoctoral awards instead of the approximately 500 that had been anticipated before the requirement of loyalty investigations was enacted. Renewals of existing fellowships in some 175 instances allowed completion of nonsecret projects that had not run their full course before the end of the academic year in the spring of 1950. For the future the fellowship program will become in essence merely an element of the researches that are carried on secretly under AEC auspices. No longer will the AEC support the broader, fructifying work of young Americans at the fast-changing frontiers of science in the atomic age. The midsummer madness that a lone Communist youth aroused in Congress has in the end caused the reorientation of the entire fellowship program. One may well conclude that the blow Congress aimed at Communists has instead left the nation a little less well equipped for the future than otherwise it might have been.

What remains as a question mark is whether the AEC fellowship controversy will prove to have been an isolated episode. For a time there was reason to believe that all who receive grants would henceforth be deemed the recipients of "handouts," to be exposed to whatever qualifying tests might please a somewhat condescending patron. Educators were fearful that college and university faculties whose salaries may be paid in part with funds received from the Government might be subjected to loyalty tests, while students whose educational costs are satisfied out of tax revenues might become objects of censorial concern lest "disloyal" youths be educated at public expense.

These fears were given great impetus by amendments to the National Science Foundation bill in the House of Representatives in 1950. One of those amendments provided that the Foundation should award no scholarship to any person "unless and until the Federal Bureau of Investigation shall have investigated the loyalty of such person and reported to the Foundation such person is loyal to the United States, believes in our system of government, and is not and has not at any time been a member of any organization declared subversive by the Attorney General. . . ." This provision was vigorously opposed by the Federal Bureau of Investigation itself, which did not wish to have the responsibility for evaluating as well as collecting evidence. It was opposed, too, by the Attorney General, the Secretary of Defense, and many others, who felt that the proposal far exceeded the necessities and would disregard American concepts of justice by penalizing past (and perhaps innocent) membership in an organization listed by the Attorney General.[44]

When the National Science Foundation bill came before Congress for final action, the offensive amendment was stricken. In its place was enacted a provision, section 15(d), that no scholarship or fellowship may be awarded to any individual unless he (1) files an affidavit that he does not believe in or support any organization that believes in the overthrow of the United States Government by force or by any other illegal or unconstitutional methods and (2) takes an oath that he "will bear true faith and allegiance to the United States of America and will support and defend the Constitution and laws of the United States against all its enemies, foreign and domestic." As to researches touching matters of AEC or military interest, of course the customary personnel security measures remained fully applicable in addition.

The contest over the National Science Foundation measure brings into true perspective the contest over the AEC fellows. The question was not whether the recipients of fellowships were a menace to the nation's security. The question was whether they could pass muster as loyal Americans.

Insistence that no youthful researcher may share in a public program unless he can be stamped as orthodox may lead too quickly to sterility. Individuality and intellectual diversity have not been flaws in the nation's structure; they have been its strength as well as its spice, and nowhere has this been more true than in its academic institutions. Freedom as we know it has not grown out of a standardized brand of Americanism; it has grown out of the political disharmony, including even the full range of political extremism, which is a central element of this country's tradition. As to the young it is perhaps especially important that independence be encouraged. The intellectual follies of youth rarely survive experience. But they have value nonetheless as indications that young people's minds are active rather than lethargic.

Authoritarianism of all sorts has been minimized in the United States, especially in educational matters, because the liberation of man's mind follows the pursuit of doubt rather than the passive absorption of dogma. Americans, having never willingly accepted official dictation of their thoughts, have always looked with disfavor upon official inquiries that could be the first step toward regimentation. All these values may be placed in issue by the further unfolding of the problem raised in the context of the AEC's small fellowship program. Unless the whole educational machinery is ultimately to become "co-ordinated" by governmental demand, investigation of character, associations, and opinion must stop at the academic gates when no vital interest demands that the gates be opened.

VIII

The Need for Fair Procedures

PROCEDURAL questions are too often dismissed by non-lawyers with an impatient shrug and the muttered belief that they are "mere technicalities" or "a lot of legalistic hair-splitting." It is unfortunately true that procedural objections are sometimes used by lawyers to becloud rather than clarify the substantial issues of disputed cases. Occasionally, too, the layman is painfully aware that a certain type of attorney enjoys jousting over technical points without regard for the merits of the controversy. Sound procedures are, nevertheless, powerful handservants of justice. Justice, to be sure, can never be produced by procedures alone. But procedural safeguards can and do minimize the opportunity for *injustice*.

Government, said William Johnson, one of the ablest members of our early Supreme Court, is "the science of experiment." Through centuries of experiment the processes of governmental fact finding have been refined and improved. By close attention to methods, government has progressively cast off the error-producing crudities that marked the investigations and trials of former days. Now, under the pressure of supposed peril and contrary to all the teachings of experience, some of those same crudities are being reintroduced into contemporary proceedings.

One hears it said from time to time that we should not today be overly concerned with fair procedures because, as some officials have put it in informal off-the-record conferences, in loyalty or security matters "it is far better that nine innocent men should suffer than that one guilty man remain unconvicted." A moment's thought should suffice to dispose of so crass a comment. Fair procedures do not prevent the detection of wrongdoing. We need not convict the innocent in order to avoid freeing the guilty. The choices before us are, rather, whether we shall preserve effective and just means of finding out who is innocent and who is not or whether, on the other hand, we shall carelessly cast a net large enough to enmesh the guilty and the innocent alike. If the latter course be pursued, is it not likely that self-respecting persons will simply shun the area in which the net is being cast?

A Fair Opportunity to Defend

Since the signing of Magna Carta in the thirteenth century, notice of charges and an opportunity to be heard before being condemned have been central characteristics of Anglo-American justice. It is vain to give a man his day in court if he has no effective opportunity to prepare for it. To commence a trial without first giving reasonably precise information concerning the matter that is to be heard is to mock the "due process of law" which tradition and constitution alike demand in this country.

Today, despite tradition and constitution, serious inquiries into a person's character and loyalty are often initiated by accusations so broad as to be virtually meaningless.

A university scientist, denied clearance by the Army–Navy–Air Force Personnel Security Board and therefore faced with termination of his appointment, requested notice of the charges against him and a chance to be heard in defense against them. The following is an exact quotation of the accusations he was called upon to meet:

"(1) That you are sympathetic with the aims of the Communist Party of the United States, its principles and practices; and

"(2) That you associate with persons who are sympathetic with the aims of the Communist Party."

An employee of a federal agency which by no stretch of the imagination had any "sensitive" work in its charge, received the following statement as an indication of why her loyalty was called into question:

"The Commission has received information that you have been sympathetically associated with Communists and/or pro-Communist individuals."

A chemist in one of the national laboratories supported by the Atomic Energy Commission received the following charges:

"It is noted that he has stated he is an ideological Communist and although he is not an active party member, or a joiner of Communist organizations, he is sympathetic to the Communist cause."

Charges like these are patently inadequate. In ordinary proceedings the vagueness of the accusations might be removed by the evidence that would ultimately have to be brought forward in their support. In matters of the sort we are discussing, however, it is not required that the charges be sustained by testimony of witnesses or by any other evidence. In a real sense the charges *are* the evidence, and the burden of overcoming them rests upon the person whose status has been questioned. This may be a well-nigh impossible task. It is not easy to prove oneself innocent of the offense of having stated to an unidentified person at an unspecified time and place that he was an "ideological Communist." It is not a simple matter to show nonassociation with "pro-Communist individuals" whose names are not revealed. It is difficult to establish one's lack of sympathy with "the aims of the Communist Party" when there is no specification of what the accuser has in mind. In practice most of the loyalty boards have decently tried to indicate in a general way the evidence that has generated the charges. But the practice varies considerably, not only from agency to agency but even from case to case within a single agency. The upshot is that in a number of instances persons have been compelled to exonerate themselves by proving a general negative, rather than by simply discrediting the evidence against them. As every lawyer knows, this is a difficult feat. It can rarely be performed without expending tremendous exertions to overcome undisclosed and possibly wholly imaginary information of an adverse character.

In fact, even after the exertions have been made in cases of the types discussed in preceding chapters, one never knows whether he has addressed himself to the evidence on which the charge was based. Let us consider in this connection an entirely nonpolitical accusation, in order to exclude any emotional or ideological overtones. Here is a "model charge" set forth in the language officially recommended by the AEC's General Manager in his instructions to the staff:

> "The Subject, according to the information obtained from responsible persons, is indiscreet in his conversations when intoxicated and has discussed in public places restricted information relative to his work."

If one suppose himself for the moment to be "The Subject," what does one do in order to meet this charge? Of course The Subject at once rallies the best available character witnesses, who will testify that he never becomes intoxicated or that, if he does, he is not indiscreet in his conversations. He offers the testimony of his local pastor to prove how highly he is regarded in his home community. All in all he makes a highly persuasive showing that he is an admirable fellow indeed. But when everything is said and done, has he disproved the charge? Of course he has not, for none of his witnesses was present at the times and places when he was said to have been indiscreet.

Why then, asks the layman, does he not produce witnesses who were present at those times? The answer is, obviously, that The Subject cannot find out when or where his misdeeds allegedly occurred. Whatever evidence there is on these points reposes in the FBI investigation report, and may not be revealed. Unless the FBI's informant agrees to be named, the fact that he furnished material is held under a pledge of secrecy. If the time, the place, the surrounding circumstances were spelled out in a charge, The Subject might be able to identify the person who had made adverse statements about him. And if this were possible, the FBI believes (and has persuaded the other agencies) that there would be great difficulty in obtaining the comments that now fill its reports.[1]

It is hard for most Americans to realize that, in proceedings of such great seriousness, decision may rest in the end upon the unrevealed testimony of undisclosed informants. This is not a matter of protecting "undercover agents" or counterspies. It is, rather, a matter of guaranteeing the anonymity of unofficial informers—social acquaintances, fellow-employees, neighbors, tradespeople, and the like, who would perhaps be embarrassed if The Subject knew what they had said about him. Yet it is precisely in this setting that the traditional American procedures of confrontation, cross examination, and rebuttal have their greatest importance. Vital as they may be to The Subject, who may otherwise be victimized by the malice or mistake of the informer or the erroneous recording of the investigator, these procedural protections are no less valuable to the body that must make a decision concerning disputed facts. They are surely among the most effective instruments known to the law for discovery of the truth. They make possible a searching of motives, accuracy, and veracity. Nowhere, perhaps, is that sort of searching more important than in cloudy disputes about loyalty and security.

General William J. Donovan, the wartime head of the O.S.S. and certainly no stranger to problems of espionage and counterespionage, has expressed belief that "much more could be disclosed to the employees than is currently disclosed." In his view of the matter, a sharp distinction should be drawn between the casual and the professional source of information. As General Donovan puts it, "There seems no reason why the anonymous informant who is not in the regular employ of the FBI and whose testimony is relied on by the Board should not be revealed to the employee . . . If non-confidential informants do not want to stand up and be counted, then their information should be used only as possible leads and not be made the basis of a record which cannot be refuted. And where it is impossible to reveal to the employee the *source* of the evidence against him, as in the case of confidential informants, the employee should at least be fully apprised of the *contents* of the testimony." [2]

In one of its recent reports the Joint Congressional Committee on Atomic Energy has discussed the case of an employee about whom derogatory remarks had been made in an anonymous letter. When the FBI followed up the letter, it obtained damaging statements from several informants. In due time charges were filed and a hearing scheduled. Then the local hearing board, acting in response to the affected employee's pleas, made an especial effort to persuade the four principal witnesses to appear. Three of them agreed to do so. One of them, who had earlier given the FBI apparently relevant information, declared under oath that "he had no basis in fact whatsoever to support" his statements. The sworn testimony of the other witnesses as well seems to have been drastically weakened, for the hearing board proceeded to clear the affected employee. The case, as the Joint Committee well said, "serves as warning that an informant may perhaps give the FBI highly unfavorable advice, but, when placed under oath before a local board, deny all that he had said, admit that he knows little or nothing about the employee, and admit further that he bore him a grudge." [3]

This does not constitute a criticism of the FBI. The impressions of casual, nonofficial informers may sometimes properly serve as leads to more conclusive evidence; there is certainly no reason why the FBI should reject "tips" even though they themselves do not constitute proof of anything. As former Attorney General Clark once stated, when embarrassed by the contents of an FBI file which had been read into evidence in a judicial proceeding, "That hearsay or gossip should appear in the investigative reports is to be expected. It is in this way that complaints and information accumulate." [4] Moreover, there is much to be said for Mr. J. Edgar Hoover's view that the FBI should record all the material it may acquire, leaving it to others to evaluate the significance of what has been recorded. The practice of indiscriminate recording, however, places an especial obligation upon fact finders to avoid drawing conclusions merely because a statement is embodied in an FBI file. One ought always to recognize the riskiness of relying upon the fallible testimony of individuals whose ability and desire to recollect and narrate truthfully have not been subjected to the test of cross examination.

In one case involving an Oak Ridge scientist, the FBI reported that six persons had told its agents that the man's wife was an active Communist. This was enough to raise a question about the scientist's suitability for continued employment. When the case came on for a hearing, the AEC asked the six witnesses to appear voluntarily to give their testimony openly, although if they had declined to do so their statements to the

FBI would presumably have been utilized in any event as "statements by confidential informants." At the hearing, three of the six said they knew of no Communist leanings, although one said that she had heard office rumors which she herself did not believe. A fourth witness testified that he had heard the employee's wife say, "The Russian government has done more for their people than the American government has done for their people." A fifth witness claimed he had seen the employee's wife at Communist meetings, but his testimony was vague and contradictory, and he was unable to identify her photograph when it was mixed with others. As the hearing progressed, the cumulatively damaging "evidence" simply vanished. The suspicions aroused by the unchecked testimony were dissipated, the affected employee was cleared, and a capable scientist was saved for the nation's undermanned laboratories.

Cases like this are no novelty in the experience of any lawyer or, indeed, of any police investigator, newspaper reporter, or business executive who has sought to ascertain the true facts amidst a mass of conflicting statements. That is why the Supreme Court believes that "judgment on issues of public moment" is likely to be treacherous if based on testimony "not subject to probing by judge and opposing counsel." [5] In the loyalty and personnel security cases, the adverse evidence is not only not subject to "probing," but much of it is actually unknown to the person against whom it is used and he therefore has no opportunity whatsoever either to discredit or rebut it. And if by chance the defendant in one of these proceedings were to guess the identity of his accusers and were to desire their presence as witnesses, nothing could be done to compel their attendance if they chose to remain away. Neither the loyalty boards nor the Industrial Employment Review Board nor the Atomic Energy Commission has been given power to issue subpoenas, to pay fees and expenses of an employee's witnesses, or otherwise to assist the development of the defendant's case. It is only fair to add, however, that a considerable number of boards have conscientiously tried to encourage the appearance of those whose testimony might be relevant. Several boards, moreover, have held hearings in different states in order to facilitate the production of witnesses.

If, as the investigating agency is convinced, disclosure of the adverse testimony is often not feasible, some other sort of protective mechanism must be developed in order to minimize injustice. As Mr. Justice Jackson said when Attorney General, investigative reports sometimes include "the statements of malicious and misinformed people." [6] We know, too, that courts view with great suspicion the testimony of informers, even when given in open court and subject to the safeguards of judicial trials.[7] We are aware that in times of political and social tension a whole community may become infected with the "informing spirit." [8] When denunciation of the citizenry becomes widespread, its reliability tends to decrease, for a sense of exactitude rarely accompanies a frenzied desire to "expose." Political talebearing feeds on the poisons of fear and suspicion. It helps create yet larger fears. Ultimately those fears serve to distort the truth, occasioning distress in the end to both the fear-ridden and the fear-victimized.

All these considerations point up the need for further procedural safeguards than now surround the use of FBI reports. In all probability the courts, while recognizing that present methods would not satisfy the constitutional requirements of due process, will hold that due process need not be afforded a federal employee in connection with his job.[9] The employee has no "property interest" in his position, and a long chain of judicial utterances suggests that the Government owes its employees no legally enforceable obligation to accord them fair treatment. This, however, is not the heart of the question.

The true issue is not whether the Constitution *forces* the Government to observe the rudimentary procedural decencies. The issue is whether the Government ought to observe them voluntarily as a matter of policy rather than because of compulsion. At present the Loyalty Order itself provides that charges need be only as specific "as, in the discretion of the employing department or agency, security considerations permit . . ." Shockingly vague charges of the sort set forth earlier in this discussion show how loosely this discretionary power has been exercised. Moreover, the FBI's insistence upon non-identification of witnesses when it is "essential to the protection of the informants or to the investigation of other cases that the identity of the informants not be revealed," seems to have been undiscriminating. Nondisclosure of the witnesses' identity has been the rule not only when concealment was "essential," but also when it would have been merely embarrassing or inconvenient to name the informants. Whether the Constitution demands them or not, fairness and moderation are the responsibility of government—a responsibility owed to public employees and to all other citizens alike.

Findings and Decisions

Arbitrariness can be minimized if care is exercised in formulating the specifications of a charge of disloyalty. Injustice can be lessened by rejecting the undisclosed testimony of unidentified witnesses. But these are not the only means of diminishing caprice and mistake.

In courts and generally in administrative agencies specific findings furnish a foundation for dispositive judgments, so that the relationship of the conclusion to the evidence may be discerned. In an appellate court, moreover, the process of reasoning leading to decision is customarily reflected in a written opinion, which is then subject to critical examination by the legal profession and the public at large. While perhaps not necessary in repetitive situations marked by only a narrow range of facts and by well-defined criteria of judgment, formal opinions are especially useful in areas where the standards of discretion are still evolving. When an adjudicator knows that he must record his judgments and give reasons for them, there are fruitful psychological effects. In Felix Frankfurter's words, we all feel much more responsible "if we have to sit down and write out why we think what we think." [10] As the Attorney General's Committee on Administrative Procedure said, "There is a salutary discipline in formulating reasons for a result, a discipline wholly absent where there is freedom to announce a naked conclusion. Error and carelessness may be squeezed out in the opinion-shaping process." [11] In proceedings that are subject to judicial review, the courts have repeatedly emphasized "the need for clarity and completeness in the basic or essential findings on which administrative orders rest." [12] Especially in cases where decision may have been based on one or more of several possible grounds, or where the play of discretion may be extensive, there is a well-settled judicial conviction that the administrative tribunal should carefully explain what its thoughts were.[13]

In the face of this sort of sentiment, the fact finders in loyalty and personnel security proceedings almost uniformly fail to make findings or prepare opinions which will illuminate their decisions. Of all the agencies that operate in this field only the Industrial Employment Review Board, through procedural regulations which became effective near the close of 1949, provides that "the decision of the Board, which shall state its reasons therefor, will be set forth in writing." Experience with this innovation is as yet too limited to permit appraisal of its effects.

The Loyalty Review Board has taken an extreme stand in support of mystification instead of explication. It has gone so far as actually to forbid the preparation of findings or explanatory statements by any of the boards whose work it superintends. This prohibitory directive assures that whatever is rational in the loyalty program will be obscured, while the irrational is fostered through being concealed.

Even the Atomic Energy Commission, which on the whole has exhibited a keener sensitivity to human values than have some of the others in this trouble-laden business, has been inattentive to the matter of findings. Its procedures contemplate that the Personnel Security Board which conducts the initial hearing will recommend to the local AEC manager that clearance should be withheld or granted, as the case may be. The manager reviews the recommended decision and transmits it with his own recommendation to the General Manager of the AEC in Washington. If the recommendation is adverse, the employee is informed of it and is given opportunity to seek review by the three-man Personnel Security Review Board. That body, which is purely advisory, then makes its recommendation to the General Manager, who, perhaps after first consulting the Commission on policy issues, takes the ultimate step of issuing or denying clearance. At no point in this long chain of recommendations and judgments is any statement given the employee as to the reasons for the decision that so vitally concerns him.

So far as the affected employee is concerned, the absence of any particularization often leaves him wholly in the dark as to the nature of his offense. In an earlier section we have noted the generality of some of the charges. If at the end of such a case no reasonably specific findings have been recorded, the employee has not only gone to hearing without having been apprised of what was to be heard, but may come away at last without ever learning what it is that supposedly warrants his loss of standing. Even in the cases in which charges are adequately detailed, he has no way of knowing which of the charges have been sustained by the hearing board; and so he is handicapped in focusing his further efforts to clear his good name. Moreover, few hearings confine themselves to the charges that were stated at the outset. Hence one cannot be sure whether the decision relates to some of the formal charges or, rather, to some other issue that was freshly introduced during the hearing.

The transcript of hearing in a case involving a nuclear scientist has been selected at random from those available and has been analyzed specifically to illustrate the intrusion of new issues into a formal proceeding on charges. Here there were but four charges. The first accusation was that the scientist had at one time read a certain left-wing publication. The other charges were to the effect that three of the man's relatives by marriage had been reported to be Communist sympathizers. During the hearing the following additional matters were extensively explored:

1. The co-operative movement in the United States and the defendant's opinions concerning either co-operative or state ownership of property;
2. The defendant's reading habits, unrelated to the particular charge;
3. The defendant's approval or disapproval of "our capitalistic form of enterprise in the United States";
4. The defendant's beliefs as to whether the country was being well-governed by the present Administration;
5. The depth of the defendant's convictions about eliminating racial injustices;
6. The allegation that one of the defendant's in-laws had been interested in "assisting members of the Abraham Lincoln Brigade in Spain in regaining American citizenship";
7. The employment record of the defendant's father and the extent of his participation in labor union affairs.

The formal charges in this case were so scantily treated during the hearing that they appear to have been deemed almost irrelevant. Once the hearing started, the signal was given for a virtually limitless inquiry. In this particular case the man was cleared. If he had not been, it would be difficult to say whether the adverse finding rested on the declared charges or whether, instead, they related to the defendant's opposition to racial discrimination, his failure to give enthusiastic endorsement to the free enterprise system, or his parent's union activity.

The hearing just reviewed is by no means exceptional. Each case typically involves an inquiry into many aspects of an individual's social and political life. The choice of what inquiries should be pursued is largely, though not exclusively, influenced by the inquisitor's hypotheses and standards, rarely articulated, often unconscious, and not invariably sound. This but emphasizes the need of candor and care in stating the real grounds of a decision, so that erroneous presuppositions and irrelevancies may be detected if they have affected final judgment.

If the boards announce their reasons as well as their results, they will be constantly reminded of the gravity of the acts they perform and will be stimulated to relate their conclusions to the evidence at hand.[14] Unexplained decisions cannot be subjected to effective analysis either by the immediately affected employee or by a public that does not desire that decency be garroted in the name of safety. Judgments that a man is "disloyal" or that his having access to restricted data will "endanger the common defense or security" ought never to be made unless they reflect rationally defensible conclusions from specific findings. Any lesser requirement leaves too great room for whim, malice, or plain stupidity.

Action on Applicants for Employment

Procedural defects of the sorts discussed in preceding sections are properly censurable. But even defective procedures are better than none at all. That is why the Atomic Energy Commission has been especially subject to criticism despite its generally moderate approach to personnel security problems. Unlike other agencies dealing with these ticklish issues, the AEC alone consistently fails to provide some sort of hearing opportunity for a person whose entry into employment is conditioned upon his obtaining clearance.

Under the Loyalty Order a "new employee," defined as a person entering a federal position after October 1, 1947, is entitled to a hearing before a regional loyalty board appointed by the Civil Service Commission if his loyalty is doubted. To be sure, a person who had been flatly rejected by an employing agency because of real or fancied question about his loyalty would probably never know why he had not been hired. Ordinarily, however, employees in the general Government service are put on the job for a probational period during which the loyalty check or investigation is made. Thus they become "new employees" with whatever procedural privileges attach to that status. Under military clearance procedures, moreover, a person whose employment in private business has been blocked by the military's Personnel Security Board is assured a hearing before the Industrial Employment Review Board. But if clearance is denied a scientist who needs an AEC clearance in order to enter upon work with classified data, he may never be able to discover why his clearance has been withheld. Nor may he challenge the soundness of the adverse decision. An AEC hearing is available only in connection with the possible revocation of a clearance that has already been granted on a prior occasion. Newcomers have no rights whatever.[15]

As a consequence, applicants for employment encounter markedly increased perils and perplexities. A scientist who has been recruited for particular work in one of the national laboratories, or who has sought employment by an AEC contractor such as the General Electric Company, may be kept in suspense for literally months while his eligibility is being judged. If doubts arise, they are not disclosed to him, nor is he afforded an opportunity to shed light upon them. Sometimes the decision is so intolerably delayed that the applicant simply moves to other jobs in order to avoid starvation. Ordinarily the financial or professional consequences of a delayed or denied clearance are as disastrous to scientist-applicants as they are to the "old hands" in the atomic energy program. The future employment opportunities of a construction worker or truck driver are not likely to be drastically lessened by his having failed to obtain an AEC clearance at some point in his life. This is not so, however, in the case of a man equipped for work as a nuclear chemist or physicist or radiobiologist. If the AEC has rejected him on security grounds, his whole professional career will be shadowed and confined. He will almost certainly find fewer and fewer chances to utilize his professional skills, for clearance requirements, as we have seen, are being steadily broadened rather than narrowed. In sum, a scientist, whether he is an applicant or a man already at work in an AEC installation, may be virtually ruined if he is held to be unemployable as a bad security risk.

What, then, are the reasons for differentiating between incumbents and applicants in respect of the procedures by which eligibility is determined?

The chief reason is not even debatably defensible. It is, in brief, that cases can be disposed of without accountability. This has been achieved in some instances by informally intimating to a would-be employer that a security *doubt* exists concerning a named applicant for whom clearance has been sought. When this is coupled with a suggestion that decision will not soon be forthcoming and that perhaps the employer may want to consider another man for the job, the hint needs no further reinforcement. In cases of this sort, the offer of employment is likely to be withdrawn, the applicant being given whatever explanation the employer deems suitable. Through this device the AEC's security officers are spared the unpleasant difficulty of actually deciding whether the available evidence warrants a denial of clearance. Reliance upon this substitute for judgment should be firmly repressed by the AEC itself. There is no justification for it, in morality or even in terms of administrative expediency. When an application has been made for permission to give a scientist access to restricted data, the professional qualifications of that scientist have been approved by the proper employing authority. Thereupon the AEC's duty becomes clear and single. The AEC should decide the issue of security clearance. The obligation to do so is evaded rather than fulfilled when AEC staff officers covertly influence the employer to withdraw the clearance application. It is in just such cases that the opportunity for ill-informed and unfair judgments is greatest, for in these instances the file may be closed without apparent administrative responsibility and without the careful findings which reinforce that responsibility.

The exact extent of the practice just discussed is, of course, unascertainable. In the nature of things it is private and unrecorded. That it is widespread is, however, unquestionable. In the whole period from January 1, 1947, to April 30, 1948, the AEC *formally* denied clearance to only 173 individuals out of the 141,469 applications for clearance that had been investigated by the FBI and evaluated by the Commission. These included casual laborers and all other classes of employees. Yet one of the officials of Brookhaven National Laboratory has asserted that during approximately the same period of time he himself knew of at least one hundred recognized atomic scientists who had failed to receive clearance and had never learned the reason.[16] A recent report of the Joint Congressional Committee on Atomic Energy lends support to the opinion, based on interviews of numerous employers and AEC officers, that the formal denial of clearance accounts for a minor fraction of the cases in which clearance is in fact withheld. The Committee, reporting on October 13, 1949, disclosed that as of that time clearance had been "formally denied" in 216 cases. In 333 additional cases, however, clearance was "denied through an expedient," the persons concerned being job applicants who were simply not hired "as a result of their FBI reports." In 874 further instances involving job applicants, the requests for clearance were withdrawn "because [the Committee says without elaboration] they had meanwhile decided to work elsewhere." [17] These figures suggest that about six out of seven cases in which doubts arise are disposed of without formal action though with the same effect as though a security decision had been made.

The other reasons for the AEC's no-hearing policy are perhaps less objectionable, though on the whole they are scarcely adequate justifications for withholding protection against prejudice or mistake. They are, in brief, that hearings are expensive and, if they become numerous, annoyingly inconvenient.

In considering these reasons, one must start with the basic proposition that the grant of a hearing is not a mere act of formal courtesy. It is, rather, a means of enabling the Commission to arrive at a just and discerning conclusion. In this respect it has as great importance for the Commission as for the applicant. Convenience is a luxury the Commission can ill afford if it results in confusion masquerading as assurance. And that is exactly what the avoidance of hearings in applicant cases does produce.

This can readily be inferred from the experience of the AEC in cases involving present staff members who, unlike the applicants, may demand that they be heard before they are stigmatized. In each of these cases a full field investigation by the FBI had been made. The investigation had disclosed information which, on its face, appeared so serious that responsible and intelligent security officers believed clearance should be withdrawn. But when those same cases went to hearing, it was almost invariably found either that the information was misleading or that its apparent implications were overborne by other pertinent facts of which the authorities had been ignorant. As was said recently by Dr. John A. Swartout, director of the Chemistry Division of Oak Ridge National Laboratory, who has sat on numerous AEC hearing boards, the reader of a security file is "overwhelmed by the succession of testimony . . . and the accumulation of information which combine to set a pattern pointing to the unreliability of the suspect"; but then one learns more about the case and discovers "how misleading such an accumulation of information can actually be." That is why, in his informed opinion, the decision is frequently against the granting of clearance "when the review is based only on the file," while reversal of an adverse decision is nearly invariable when there has been a chance to meet the seemingly damning evidence.[18]

It would be silly to suppose that a hearing for an applicant would not be equally likely to dissipate the objections to him. The fact is that the security file can rarely tell the full story about any man. Although every word it contains be unassailably accurate, the file is yet unlikely to give a rounded impression of the subject. This is not because the FBI or anyone else has suppressed evidence that is favorable. On the contrary, Mr. Hoover has emphatically instructed his agents that they are not conducting an investigation for the purpose of establishing doubts. The object of the inquiry is to acquire full knowledge, the good along with the bad, about a man. It nevertheless remains true that the primary reason for having an investigation at all is to discover whether any information exists that might be deemed derogatory; and though the investigator also reports whatever favorable comments may come to his notice, the drive of his inquiries is inevitably in the direction of what is adverse rather than commendatory.

Furthermore, even when there is the best will in the world on the part of both the information giver and the information recorder, there is always room for serious defect in an investigation report. Consider the matter of mistaken identity simply as one example of possible error. The former deputy administrator of the Office of Scientific Research and Development has asserted that a major personnel security problem during the war "arose from cases of mistaken identity in which the person wanted by the National Defense Research Council happened to bear the same name as some other person whose record was not such as to endear him either to the military service or to any other organization interested in honest operations." [19] The frequency of this problem of mistaken identity is evidenced by the fact that out of the first 7,667 full field investigations that were conducted under the Loyalty Order, 494 were discontinued because the derogatory information appearing in the files proved to have related to someone else.[20] The FBI, as Director Hoover recently wrote with justifiable pride, often successfully demonstrates that information furnished it in connection with loyalty inquiries "is incorrect or the wrong person is involved," so that exoneration instead of conviction flows from the FBI's work.[21] It would be almost miraculous if this sort of self-detection of error occurred every single time error was present.

Finally, there is rarely a security file which is so clear that no judgment is required for its evaluation. When a possibly incomplete record is read in such a way that all the possible inferences are made to fall in a single direction, the likelihood of distorted (even though conscientious) judgment is very considerable. Still fresh in public memory is the disclosure that during the summer of 1949 Army officers had declared to be "unemployable," presumably because "disloyal," a large group of distinguished Americans including Gordon R. Clapp, chairman of the Tennessee Valley Authority, Professor George S. Counts of Columbia University, Roger N. Baldwin of the American Civil Liberties Union, and many other educators and authorities whose services were desired by our occupation authorities in Germany. When the matter came to light through enterprising journalists, the Army's top officials were as amazed as they were embarrassed.[22] No one sought to justify the absurdity of so characterizing an important public administrator who had been confirmed in his post by the Senate after a searching inquiry, a prominent teacher whose record as an anti-Communist was certainly no secret, and a civil libertarian who had but recently completed surveys in Japan, Korea, and Germany at the personal request of the American military governors of those countries.

In cases of this sort one can be made readily aware of the crudities and the misapprehensions that sometimes flow from reading a loyalty or security dossier. In cases that involve less conspicuous personalities, however, there is little likelihood that a lapse of intelligent judgment will ever be recognized. Most of the scientists for whom AEC clearance is sought are not great men whose position in the community is so assured that a denial of their clearance would be greeted with derisive hoots. They are, in the main, young men of competence but not yet of note. When clearance is withheld from one of them by an AEC security official's say-so after examination of the file, the decision may be as debatable as was the Army's. Unlike the Army's, it is not likely to attract public attention. Since it deals only with a "routine case," it will remain unnoticed, uncorrected, and unconscionably hurtful to all concerned.

An opportunity to challenge a judgment of rejection is a vital necessity if needless damage is to be avoided.[28] The AEC cannot afford to wreck and abandon any of the precious resources of scientific personnel available for its enterprises.

IX

Concluding Thoughts

NATIONAL, ideological, and economic rivalries have created a modern age in which crisis is commonplace rather than phenomenal. The hazards of an unquiet world cannot be ignored. Awareness of those hazards is reflected in the policies and procedures this book has described. They are intended as countermeasures against danger. One can have small quarrel with their basic purposes.

Not every one of the actions taken in the name of national safety has, however, been praiseworthy. Some have been superfluous. Some have produced too little advantage at too high cost. Some have been crudely developed.

On the whole our national policies about secrecy in scientific matters are intelligently formulated. But the policies seem too inflexibly applied. In part this inflexibility is a product of popular pressures. These pressures reflect misconceptions about the nature of scientific knowledge, coupled with a grossly mistaken belief that the intelligence which creates new understanding is largely if not exclusively concentrated in the United States. The effort to "keep scientific secrets" is explicable as a military expedient, but it can never be wholly successful. All history demonstrates that problems solved by the laboratories of one country ultimately yield to research in others, so that permanent bottling up of "secrets" is a virtual impossibility.

Nevertheless it is feasible and sometimes desirable to restrain the dissemination of scientific learning for short periods of time. Public authorities in the United States do this today by "restricting" or "classifying" information, so that its unauthorized transmittal becomes not only a breach of professional trust but a crime against the state. The justification for this enforcement of secrecy is, of course, that possible enemies will thus be hampered in the development of their military or industrial resources. The trouble with insistence upon silence is that it is so likely to be overemphatic. Then its debatable virtues are quickly offset by heavy costs.

In the first place it is worth remembering that the progress that produces our "secrets" has always depended upon free exchange of scientific insights. We can expect to gain equally in the future from the work done elsewhere, unless we shut ourselves off from all the world lest the world learn from us. Then, too, it is well to recall the important ways in which scientific developments have implications unforeseen and unforeseeable. When German physicists demonstrated the fissionability of uranium, they were not thinking about atomic bombs. Similarly, a body of knowledge that has immediate military interest may have its most valuable, though entirely unpredictable, uses in quite unrelated settings. That is why "compartmentalization" of research has never been found efficient. The work done by one scientific investigator may have tremendous urgency for some wholly separate research project. When barriers are erected that block off one researcher from another, scientists are prevented from exchanging their learning in traditional ways and consequently everyone's rate of advance is slowed. Fragmentation of knowledge makes for inefficient training, and for imperfect utilization of available manpower. It discourages adequate application of work already done. It causes unnecessary duplication of unfruitful research. It forestalls the constructively critical advice of fellow scientists. And all these consequences are felt outside as well as within the particular areas of science in which secrecy is sought to be maintained.

Further, the constant stress on security, with the attendant discouragement of scientific interchanges, is having an intangible but real psychological effect on American scientists. A prevalent hesitancy to communicate with one's professional colleagues flows from uncertainty as to whether one's words may unwittingly reveal "secrets." This hesitancy is marked not only among those who work directly in military science, but also among teachers where its consequences are especially hurtful to the nation. Because of artificial limitations upon what can be taught, students are all too often given imperfect training in subjects that must be mastered if the next scientific generation is to build successfully on the foundations now being laid.

Perhaps in the field of secrecy excessive weight has been given to short-run values. Suppressing knowledge of American discoveries and developments may, it is true, hamper our adversaries. But suppression retards our own progress, too. And since at the moment no other country can match America's trained manpower and material resources, failure to utilize fully what we learn handicaps us probably more than it does our competitors.

Even were this not so, the case for sternly enforced secrecy would be unclear. The issue cannot be properly evaluated if only the short-run aspects are considered. Today the world's great nations live tensely, but they live at least formally in peace. While the risk of war is tragically real, there is no certainty that an explosion will occur soon or ever. If every phase of our national existence were to be ordered as though we were already engaged in a total war, or would be so engaged by tomorrow at the latest, the existence would be a grim one indeed. The welfare of the humans who populate the United States deserves advancement by instrumentalities other than munitions. There is real peril, in terms of the nation's future, in focussing solely upon the possible military implications of scientific or technological advances while ignoring their power for good in millions of civilian lives.

In sum, unless secrecy is to be permitted to choke advance, it must be cautiously invoked and then only as to matters of solely military significance. Information that has value for further general research or that can enhance the economic or physical well being of the world by being given peaceful applications ought not be buried. It ought not be buried even if we can see that somehow it might conceivably be pieced together with other bits of information to the benefit of a potential enemy in wartime. For Safety does not lie in Secrecy. It lies in the purposeful utilization, stimulation, and encouragement of the nation's intellectual resources.

Putting to one side the question of managing scientific data and turning for a moment to the management of scientific personnel, we find a somewhat parallel overstressing of concern about dangers and an understressing of concern about the humans whose services we need. Of course there is a proper place for precautionary measures. Sifting of some sort is entirely reasonable before one selects the staffs to whom important undertakings are entrusted. The only valid issue is whether protective actions have been so imperfectly designed that they jeopardize the democratic values they are meant to safeguard.

In this connection a differentiation must be made between the so-called personnel security cases and the so-called loyalty cases. The former involve scientists who must be "cleared" because their work requires them to have access to data within the zone of secrecy. Most of them are not employees of the

United States, but are on the staffs of universities or of industrial corporations having contractual relations with a governmental agency. Until recently, on the other hand, the loyalty cases affected federal employees exclusively; of late the requirement of loyalty testing has been extended by Congressional demand to students whose sole connection with the Government is that they have been awarded fellowships to further their studies. The loyalty cases, unlike the personnel security cases, involve no problem of "security risk," because those whose loyalty is in question have no contact with secrets.

In both these categories the chief inquiry has been into the ideas or associations of the scientists involved. Few if any cases have involved conduct or, even, character. By procedures far from polished, unquestionably competent scientists have been summoned to answer neighborhood gossip, to explain isolated acts of kindliness, to divorce themselves from the political attitudes of any of their relatives or other associates who happen to be "left wingers," and, in short, to establish their Americanism by proving that they are just like everybody else. Because some are unwilling to subject themselves or their families to inquisitions into their supposed opinions rather than their observable conduct, American scientific programs are often denied the services of high-spirited and badly needed men.

More important than the immediate loss of these talents, however, is the gradual acceptance of a political litmus paper test as a proper measurement of a scientist's qualifications, even when his work is wholly unrelated to confidential affairs. There is grave risk in judging men by their beliefs rather than by their behavior and their professional competence. In other countries there has been a discernible relationship between political eligibility tests and the decline of scientific achievement. There is no reason to suppose that, over a period of time, this country's experience would be any happier in that respect. Without either minimizing the demands of national security or magnifying the perils of our present course one can soberly urge re-examination of the measures now enforced.

In summary terms, the best course would be to shift the emphasis from "loyalty" as an abstraction, and to place it instead on "security." Whenever a position is "sensitive" in the sense that an incumbent will gain access to confidential matters of military or international concern, the probity of the incumbent must be assured; and in this context an inquiry into attitudes and associations may conceivably have relevance. But in any event the number and scope of investigations into these matters should be limited to the fullest possible degree. For the balance—the great bulk of the cases in which searching probes are commonly being made into what a man thinks or reads or whom he knows, rather than into what he does—larger reliance should be placed on administrative supervision than on political detection.

The danger to freedom which inheres in the present emphases of the personnel programs is, in a sense, not immediate. Although literally millions of persons have been subjected to suitability tests in which complete orthodoxy has been a guarantor of success, they yet constitute a minor fraction of the whole population. Moreover, it would be a mistake to suppose that only the orthodox have passed the tests. The unorthodox are as a rule found to be acceptable, but only after a travail their more conventional brothers have been spared. Still, the very fact that there is this difference in experience may have a large social significance. The nation's identification of conformity as a prime ingredient of reliability must ultimately discourage the acquisition and discussion of new ideas.

Now, obviously enough, there is nothing intrinsically valuable in new and "radical" ideas. More often than not they fall into the oblivion they deserve because they are overborne by the solid facts of experience. Nevertheless the ferment of new ideas is a tonic in any community. Many a novelty which was scoffed at yesterday has managed to survive, to become the dogma of today. The generating forces by which minority sentiment of one kind or another has changed into majority acceptance have been the life forces of American democracy. This country has constantly been altered without being shattered. It has not suffered the violent changes which unyielding rigidities sooner or later produce. Instead, it has been preserved by evolutionary gradualism.

If protest and criticism had been stilled, social evolution would of course have been impossible. It is only through awareness of defects that improvements come to pass. So it is that the detractor, the dissenter, the reformer has played a centrally important role throughout our history. Time after time some "troublemaker's" dissatisfaction with things as they happened to be, has drawn attention to problems which might otherwise have become magnified through being too long ignored. His has been the opinion that bit by bit became public opinion until, more often than not, it was no longer recognizably his at all.

There is nothing novel about the fact that holders of dissident opinions are not as a rule the most popular figures on the American scene. What is new is that their unpopularity is in a sense governmentally recognized through the proceedings we have been discussing. This is no boon to the United States. Every society that stilled protest by compulsion or fear has suffered immobilization and ultimate decay. That is why it would be perilous to enforce a concept of loyalty which substantially equates it with approval of, or at least non-opposition to, the political, economic, and social practices which at any given moment are dominant in the country.

Neither the loyalty program nor present personnel security procedures were meant to embody any such view. In operation, however, they inescapably do so to a considerable degree, for it is the dissidents rather than the conformers who get in trouble—and everyone except the most sturdily convinced or the psychotic runs away from trouble if he has a chance. The programs are candidly directed at Communists, who are regarded as the disciplined tools of a foreign power. But the inquiries the Government pursues go far wide of their mark. Effectively if unintentionally, the focus upon opinion as a measure of loyalty tends to discourage the holding of any opinion at all.

No scientist who has confined his interests to his laboratory, his flower garden, and his golf game has been touched by scandal. In the main those to whom the Government has brought distressing embarrassment were ones who became concerned, in a perfectly legal way, about racial discriminations or the Franco government or the importance of peaceful relations with the Soviet Union. Knowing this fact, many people now avoid the areas of nonprofessional debatability lest they jeopardize their professional futures.

If individuals were unrestrainedly to talk and organize together concerning the issues of the day, they might of course propagate many a badly mistaken idea. They might well be victimized, as others have been victimized, by persons who slyly play on honest emotions for political purposes. They might create unsettling and unnecessary doubts in place of a desirable certitude. But these are the normal wastages of the democratic process, the cost of encouraging free men to be boldly inquisitive concerning the problems of their times.

Those who devised the programs were not evilly disposed toward the great tradition of freedom in the United States. They may, however, have been ill-advised. "Struggles to coerce uniformity of sentiment in support of some end thought essential to their time and country have been waged by many good as well as by evil men," the Supreme Court said in an opinion by Mr. Justice Jackson in 1943. But efforts to discourage dissent have rarely succeeded. "As first and moderate methods to attain unity have failed, those bent on its accomplishment must resort to an ever-increasing severity. . . . Those who begin coercive elimination of dissent soon find themselves exterminating dissenters." Our whole constitutional order was designed to avoid that end by preventing that beginning. The freedom to differ—which is assuredly among the sharpest of the distinctions between the United States and the totalitarian states—"is not limited to things that do not matter much. That would be a mere shadow of freedom. The test of its substance is the right to differ as to things that touch the heart of the existing order." [1]

Does this observation skirt the real issue? Is it not arguable that the impact of our safety policies upon unpopular persuasions is merely incidental, while their real thrust is against international conspirators who masquerade as honest men? Of course that is the policies' intended direction. The difference between aim and effect is a consequence of inquiring into the beliefs and sympathies of vast numbers of individuals, on the wholly unsubstantiated theory that unsound opinion is the equivalent of unsound conduct, advocacy the equivalent of action. This is the defective core of the programs as now framed and administered.

In times like the present it is not comfortable to advise the alteration of programs that have as their declared goal the confusion of the nation's enemies. But in the field of science, as these chapters have sought to show, the loyalty and security programs have made only small and highly debatable advances toward the goal. Such as those advances were, they have been gained too dearly. It will require a high degree of personal and political courage for public figures to acknowledge the facts and now propose fundamentally remedial steps. Those who insist that shaky procedures and speculative findings, injustice

233

and hardship, are not the tools with which to build security, are likely to be misrepresented and denounced. Courageous men have, however, acknowledged error in the past. Courageous men will do so in the future. They may say, as did a prominent Bostonian in 1692 when Massachusetts was in the grip of a panic of an intensity which dwarfs our current disquietude, "It is irksome and disagreeable to go back when a man's doing so is an implication that he has been walking in the wrong path; however, nothing is more honorable than, upon due conviction, to retract and undo (so far as may be) what has been amiss and irregular." [2]

Appendix A

Declassification Policy

THE following lists of topics indicate the general present content of "Unclassified Areas" (work that can be conducted and published without prior AEC clearance), "Declassifiable Information" (data that must be officially declassified before release for general publication), and "Classified Information" (restricted data that will not be cleared for general publication). The lists are merely indicative, rather than precise statements. They are drawn from the AEC's *Fifth Semiannual Report* (1949), pp. 108–109.

Unclassified Areas

In general, item (a), the unclassified areas, covers the pure science related to atomic energy but not plant processes or specific experimental data of vital project importance. It includes:

(1) Pure and applied mathematics, except that applying to specific classified projects.

(2) Theoretical physics (except the theory of fission, of reactors, and of neutron diffusion, and weapon physics).

(3) All physical (except nuclear) properties of all ele-

ments of atomic number less than 90. Nuclear properties of most isotopes.

(4) The basic chemistry of all elements (except for the analytical procedures and technology of the production of fissionable materials) and the physical metallurgy of all elements of atomic number less than 83.

(5) Instrumentation, including circuits, counters, ionization and cloud chambers, neutron detectors (excluding fission chambers), electronuclear accelerators, such as cyclotrons, betatrons, Van de Graaff generators, etc.

(6) Medical and biological research and health studies (excluding work with elements of atomic number 90 and above).

(7) Chemistry and technology of fluorine compounds (except the specific applications in AEC installations).

Declassifiable Information

Item (b), the declassifiable information which may be expected to be found in the general literature after official declassification, includes:

(1) Most reactor and neutron diffusion theory, except for those parts involving semiempirical methods or related to specific assemblies.

(2) Certain physical properties of isotopes of elements of atomic number greater than 90, and the nuclear properties (except for certain neutron and fission characteristics) of isotopes of elements greater than 90.

(3) Analytical procedures (except for production applications); most physical and process metallurgy of elements of atomic number greater than 90.

(4) Medical and biological research and health studies with elements of atomic number 90 and above.

(5) Certain properties of experimental reactors, such as: fluxes, neutron distribution not revealing lattices

and information regarding thermal columns, and the velocity spectrum in the thermal column.

Classified Information

The types of information covered by item (c) are clearly classified information:

(1) Information on the production of fissionable material—equipment used, technology, handling, and disposition—including the technology of production of feed materials—and specifically all quantitative and qualitative output data.

(2) The technology of production and power reactors, including design, operating characteristics, and working materials.

(3) Information dealing with nuclear weapons and their components, including production technology, handling, disposition, testing, and technical data relating to military employment.

(4) Certain information relating to the operations and facilities of the United States atomic energy program which may be of value to an enemy in sabotage planning, or in studies of the strategic vulnerability of the United States or defense potential of the United States with respect to atomic weapons.

Appendix B

AEC Criteria for Determining Eligibility
for Personnel Security Clearance
(January 5, 1949)

THE United States Atomic Energy Commission has adopted basic criteria for the guidance of the responsible officers of the Commission in determining eligibility for personnel security clearance. These criteria are subject to continuing review, and may be revised from time to time in order to insure the most effective application of policies designed to maintain the security of the project in a manner consistent with traditional American concepts of justice and rights of citizenship.

The Commission is revising its hearing procedure entitled "Interim Procedure" for the review of cases of denial of security clearance and for the conduct of hearings for employees desiring such review. The Interim Procedure announced April 15, 1948, places considerable responsibility on the Managers of Operations and it is to provide uniform standards for their use that the Commission has adopted the criteria described herein.

Under the Atomic Energy Act of 1946, it is the responsibility of the Atomic Energy Commission to determine whether the common defense and security will be endangered by granting security clearance to individuals either employed by the Commission or permitted access to restricted data. As an administrative precaution, the Commission also requires that at certain locations there be a local investigation, or check on individuals employed by contractors on work not involving access to restricted data (Commission authorization to be so employed is termed "security approval").

Under the Act the Federal Bureau of Investigation has the responsibility for making an investigation and report to the Commission on the character, associations and loyalty of such individuals. In determining any individual's eligibility for security clearance other information available to the Commission should also be considered, such as whether the individual will have direct access to restricted data, or work in proximity to exclusion areas, his past association with the atomic energy program, and the nature of the job he is expected to perform. The facts of each case must be carefully weighed and determination made in the light of all the information presented whether favorable or unfavorable. The judgment of responsible persons as to the integrity of the individuals should be considered. The decision as to security clearance is an over-all, common-sense judgment, made after consideration of all the relevant information, as to whether or not there is risk that the granting of security clearance would endanger the national defense or security. If it is determined that the common defense and national security will not be endangered, security clearance will be granted; otherwise, security clearance will be denied.

Cases must be carefully weighed in the light of all the information, and a determination must be reached which gives due recognition to the favorable as well as unfavorable information concerning the individual and which balances the cost to the program of not having his services against any possible risks involved. In making such practical determination, the mature viewpoint and responsible judgment of Commission staff members, and of the contractor concerned are available for consideration by the Manager of Operations.

To assist in making these determinations, on the basis of all the information in a particular case, there are set forth below a number of specific types of derogatory information. The list is not exhaustive, but it contains the principal types of derogatory information which indicate a security risk. It will be observed that the criteria are divided into two groups, Category (A) and Category (B).

Category (A) includes those classes of derogatory information which establish a presumption of security risk. In cases falling under this category, the Manager of Operations has the alternative of denying clearance or referring the case to the Director of Security in Washington.

Category (B) includes those classes of derogatory information where the extent of activities, the attitudes or convictions of the individual must be weighed in determining whether a presumption of risk exists. In these cases the Manager of Operations may grant or deny clearances; or he may refer such cases to the Director of Security in Washington.

CATEGORY (A)

Category (A) includes those cases in which there are grounds sufficient to establish a reasonable belief that the individual or his spouse has:

1. Committed or attempted to commit, or aided or abetted another who committed or attempted to commit any act of sabotage, espionage, treason or sedition;

2. Established an association with espionage agents of a foreign nation; with individuals reliably reported as suspected of espionage; with representatives of foreign nations whose in-

terests may be inimical to the interests of the United States. (Ordinarily this would not include chance or casual meetings; nor contacts limited to normal business or official relations.)

3. Held membership in or joined any organization which has been declared to be subversive by the Attorney General, provided the individual did not withdraw from such membership when the organization was so identified, or otherwise establish his rejection of its subversive aims; or, prior to the declaration by the Attorney General, participated in the activities of such an organization in a capacity where he should reasonably have had knowledge as to the subversive aims or purposes of the organization;

4. Publicly or privately advocated revolution by force or violence to alter the constitutional form of Government of the United States.

Category (A) also includes those cases in which there are grounds sufficient to establish a reasonable belief that the individual has:

5. Deliberately omitted significant information from or falsified a Personnel Security Questionnaire or Personal History Statement. In many cases, it may be fair to conclude that such omission or falsification was deliberate if the information omitted or misrepresented is unfavorable to the individual;

6. Violated or disregarded security regulations to a degree which would endanger the common defense or national security;

7. Been adjudged insane, been legally committed to an insane asylum, or treated for serious mental or neurological disorder, without evidence of cure;

8. Been convicted of felonies indicating habitual criminal tendencies;

9. Been or who is addicted to the use of alcohol or drugs habitually and to excess, without adequate evidence of rehabilitation.

CATEGORY (B)

Category (B) includes those cases in which there are grounds sufficient to establish a reasonable belief that with respect to the individual or his spouse there is:

1. Sympathetic interest in totalitarian, fascist, communist, or other subversive political ideologies;

2. A sympathetic association established with members of the Communist Party; or with leading members of any organization which has been declared to be subversive by the Attorney General. (Ordinarily this would not include chance or casual meetings, nor contacts limited to normal business or official relations.)

3. Identification with an organization established as a front for otherwise subversive groups or interests when the personal views of the individual are sympathetic to or coincide with subversive "lines";

4. Identification with an organization known to be infiltrated with members of subversive groups when there is also information as to other activities of the individual which establishes the probability that he may be a part of or sympathetic to the infiltrating element, or when he has personal views which are sympathetic to or coincide with subversive "lines";

5. Residence of the individual's spouse, parent(s), brother(s), sister(s), or offspring in a nation whose interests may be inimical to the interests of the United States, or in satellites or occupied areas thereof, when the personal views or activities of the individual subject of investigation are sympathetic to or coincide with subversive "lines" (to be evaluated in the light of the risk that pressure applied through such close

relatives could force the individual to reveal sensitive information or perform an act of sabotage);

6. Close continuing association with individuals (friends,
relatives or other associates) who have subversive interests and
associations as defined in any of the foregoing types of derogatory information. A close continuing association may be
deemed to exist if:

(1) Subject lives at the same premises with such indi-
    vidual;
(2) Subject visits such individual frequently;
(3) Subject communicates frequently with such indi-
    vidual by any means.

7. Association where the individuals have enjoyed a very
close, continuing association such as is described above for
some period of time, and then have been separated by distance; provided the circumstances indicate that a renewal of
contact is probable;

Category (B) also includes those cases in which there are
grounds sufficient to establish a reasonable belief that with
respect to the individual there is:

8. Conscientious objection to service in the Armed Forces
during time of war, when such objections cannot be clearly
shown to be due to religious convictions;

9. Manifest tendencies demonstrating unreliability or inability to keep important matters confidential; wilful or gross
carelessness in revealing or disclosing to any unauthorized
person restricted data or other classified matter pertaining
either to projects of the Atomic Energy Commission or of any
other governmental agency; abuse of trust, dishonesty; or homosexuality.

While security clearance would ordinarily be denied in
each of the foregoing categories (A), and (B), security approval,

as distinguished from security clearance, might be warranted in those types of derogatory information mentioned under Category (B) above.

The categories outlined hereinabove contain the criteria which will be applied in determining whether information disclosed in investigation reports shall be regarded as substantially derogatory. Determination that there is such information in the case of an individual establishes doubt as to his eligibility for security clearance.

The criteria outlined hereinabove are intended to serve as aids to the Manager of Operations in resolving his responsibility in the determination of an individual's eligibility for security clearance. While there must necessarily be an adherence to such criteria, the Manager of Operations is not limited thereto, nor precluded in exercising his judgment that information or facts in a case under his cognizance are derogatory although at variance with, or outside the scope of the stated categories. The Manager of Operations upon whom the responsibility rests for the granting or denial of security clearance, and for recommendation in cases referred to the Director of Security, should bear in mind at all times, that his action must be consistent with the common defense and national security.

Notes

Chapter I

1. Dean Ridenour's comment on secrets was made in *Hearings before the Senate Special Committee on Atomic Energy,* 79th Cong., 1st Sess. (1945), p. 537; also printed under title "Secrecy in Science," 1 *Bulletin of the Atomic Scientists,* No. 6, p. 3 (1946).

2. Senator McMahon stated the problem of secrecy in relation to Congressional duties in an address before the Economic Club of Detroit, January 31, 1949.

3. The small number of those who know about atomic production figures was brought out by former AEC Chairman Lilienthal in *Hearings before the Joint Congressional Committee on Atomic Energy,* 81st Cong., 1st Sess. (February 2, 1949), p. 6.

4. The prevailing attitude toward discussion of atomic energy problems is well described in Anne W. Marks, "Washington Notes," 5 *Bulletin of the Atomic Scientists* 158 (1949).

5. J. R. Newman and B. S. Miller, in *The Control of Atomic Energy* (McGraw-Hill, 1948), pp. 179–180, recapitulate the testimony of leading American scientists concerning our past dependence on basic research conducted in other countries.

6. The comment on the foreign origin of leading American scientists is by S. A. Goudsmit, *ALSOS* (Henry Schuman, 1947), pp. 238–239.

7. The comment on the resonant cavity magnetron appears in James Phinney Baxter III, *Scientists against Time* (Little, Brown & Co., 1946), pp. 141–142. Additional discussion of important contributions to us by British and Canadian scientists appears in the same volume at pp. 119 ff. The contributions of British scientists to war researches that are sometimes regarded in this country as "100 percent American" are well summarized by J. G. Crowther and R. Whiddington, *Science at War* (Philosophical Library, 1948). Incidentally, Karl T. Compton, after acknowledging that the English magnetron tube made our radar possible, has written that in postwar Japan he "saw an essentially similar magnetron tube which had been described in publication by the Japanese even

earlier" than the British discovery. Surely this is a striking illustration of the international distribution of scientific talent. K. T. Compton, "Science Fears an Iron Curtain," 36 *Nation's Business*, No. 6, pp. 47, 60 (1948).

8. In connection with German attitudes toward scientific supremacy, it has recently been said by Frederick Seitz, professor of physics at the University of Illinois, "In 1940 the Germans possessed air supremacy on the basis of developments which had taken place five years earlier. At that time they were completely confident that their accumulation of talents was so unique that it would be essentially impossible for any other nation to match them in the near future, let alone outstrip them. Yet that is precisely what the United States did in a remarkably short time." F. Seitz, "The Danger Ahead," 5 *Bulletin of the Atomic Scientists* 266 (1949).

9. The discussion of coincidence in science draws in part upon a statement of Dean Ridenour in *Hearings before the Senate Special Committee on Atomic Energy*, 79th Cong., 1st Sess. (1945), pp. 537–538; also in "Secrecy in Science," 1 *Bulletin of the Atomic Scientists*, No. 6, p. 3 (1946). Further illustrative material may be found in Lancelot Law Whyte, "Simultaneous Discovery," 200 *Harper's Magazine* 23 (February 1950).

10. Dean Ridenour, in the statement cited in note 9 above, asserts that Veksler had suggested an accelerator of the synchrotron type fully two years before it was developed in this country. He adds: "The synchrotron involves a magnet, whose design is straightforward but complicated. McMillan is presently (1945) building a synchrotron, on funds supplied by the Manhattan District [the Army-administered atomic energy project which was the precursor of the Atomic Energy Commission]. When a physicist at M.I.T., who is also planning the construction of a machine of this type, asked McMillan for his magnet design, he was told that the Army would not permit the release of information on the magnet. Whom are we attempting to handicap by such restrictions? Surely not the Russians; they not only invented the synchrotron, they did it earlier than we did."

11. For fuller discussion of the isolation of bacterial toxin, see Theodor Rosebury, *Peace or Pestilence* (Whittlesey House, 1949), pp. 77–78, 188; the papers in question are C. Lamanna, O. E. McElroy, and H. W. Eklund, "The Purification and Crystallization of *Clostridium botulinum* Type A Toxin," 103 *Science* 613 (1946), and L. Pillemer, R. Wittler, and D. B. Grossberg, "The Isolation and Crystallization of Tetanal Toxin," 103 *Science* 615 (1946).

12. Discussion of Russian rebuffs of American overtures may be found in *Cultural Relations between the United States and the Soviet Union*, U.S. Department of State Publication, 3480 (April 1949), especially at pp. 2, 5, 10, 16, 17, 18–19.

13. The summaries of scientists' conflicting opinions concerning secrecy policy are derived from *Administration for Research* (Vol. III of *Science and Public Policy, A Report to the President*) by John R. Steelman, chairman, the President's Scientific Research Board (October 4, 1947), pp. 34–37.

14. The voluntary imposition of publication restraints in connection with prewar nuclear energy research is well described by H. H. Goldsmith, "The Literature of Atomic Energy of the Past Decade," 68 *Scientific Monthly* 291 (1949).

15. The voluntarily suppressed report on germ warfare is entitled "Bacterial Warfare: A Critical Analysis of the Available Agents, Their Possible Military Applications, and the Means for Protection against Them," by Theodor Rosebury, E. A. Kabat, and M. H. Boldt. It was submitted to the National Research Council in the fore part of 1942 but was not published until May 1947, when it appeared in 56 *Journal of Immunology* 7.

16. The prohibitions against communicating "restricted data" may possibly apply (and penalties may possibly attach) even to communication of relevant data that have been acquired through independent research and wholly without relationship to the official operations of the United States in the field of atomic energy. See J. R. Newman and B. S. Miller, *The Control of Atomic Energy* (McGraw-Hill, 1948), pp. 216 ff. Compare H. S. Marks, "The Atomic Energy Act: Public Administration without Public Debate," 15 *University of Chicago Law Review* 839, 845 (1948).

17. Dr. Oppenheimer commented on radioisotopes in the hearing before the Joint Committee on Atomic Energy, *Investigation into the United States Atomic Energy Project*, 81st Cong., 1st Sess. (June 13, 1949), p. 284.

18. The problems of the AEC's isotope program are discussed in the report of the Joint Congressional Committee on Atomic Energy, *Investigation into the United States Atomic Energy Commission*, 81st Cong., 1st Sess., Senate Rep. No. 1169 (1949), pp. 42–47.

19. The figures on AEC classification decisions are derived from statistics in Appendix 6 of the AEC's *Fifth Semiannual Report* (1949), p. 180.

20. Wartime classification policy as to medical research is described in Irvin Stewart, *Organizing Scientific Research for War: The Administrative History of the Office of Scientific Research and Development* (Little, Brown & Co., 1948), pp. 290–291.

21. The comment on failure quickly to declassify medical research is in L. N. Ridenour, "Secrecy in Science," 1 *Bulletin of the Atomic Scientists*, No. 6, pp. 3, 8 (1946).

22. The AEC's *Sixth Semiannual Report* (July 1949) is devoted largely to a discussion of the Commission's work in biology and medicine. A sampling of the AEC's declassification work, covering January 1950, shows that in that month 69 reports of work in atomic energy laboratories were abstracted and made available upon request. These reports fell into five classifications: Biology and Medicine, 16 reports (226 pages); Chemistry, 23 reports (680 pages); Engineering, 1 report (12 pages); Mineralogy, Metallurgy, and Ceramics, 5 reports (56 pages); and Physics, 24 reports (490 pages). See AEC Release No. 267, March 12, 1950.

23. The quotation from the *Report to the U.S. Atomic Energy Commission*

by the Industrial Advisory Group appears in 71 *Mechanical Engineering* 205, 208 (1949), and in 5 *Bulletin of the Atomic Scientists* 51, 53 (1949). Impressed by this report, the AEC appointed a working party of representatives of technical and engineering societies and the business press to commence in early 1950 an examination of technological information in AEC files bearing upon metallurgy, with a view to determining its possible value to American industry. The study was planned as a trial program to determine how much material of special interest to industry is still classified but potentially declassifiable. See AEC Release No. 239, December 28, 1949.

24. A summary statement of the areas that the AEC now denominates as "Unclassified," "Declassifiable," and "Classified" appears in Appendix A, at p. 235 of this volume.

25. *Report to the U.S. Atomic Energy Commission* by the Industrial Advisory Group, in 71 *Mechanical Engineering* 205, 207, 212 (March 1949).

26. The work of the technological working party set up by the AEC in response to the Industrial Advisory Group's recommendation is described in AEC Release No. 281, April 25, 1950.

The listing of patents available for licensing on a nonexclusive and royalty-free basis is reflected in AEC Releases Nos. 261, 279, 283, and 294, Feb. 24, April 21, May 8, June 27, 1950. As of July 1, 1950, a total of 138 Commission-held patents had been released; and more than half of these had been listed within the immediately preceding five months. In addition the Commission occasionally made separate announcements of developments of commercial or industrial interest, e.g., AEC Release No. 274, March 29, 1950, announcing the development of a new inexpensive paperlike filter material designed originally for filtering fine radioactive particles from contaminated wastes, but apparently useful also in many types of industrial filtering.

The release of data about low-power reactors, electromagnetic separation, and other wartime processes now more or less obsolete, had not officially been announced as this book went to press. It was foreshadowed, however, in utterances by AEC members (e.g., by Commissioner Gordon Dean in an address before the Blue Pencil Club of Ohio, Columbus, Ohio, May 28, 1950) and in press dispatches that were informally confirmed to me by AEC officials. See Anthony Leviero's reports to the *New York Times*, June 15, 1950, p. 1, col. 8, and June 25, 1950, p. 10E, col. 1.

27. The statutory base of classification of documents by the several departments is to be found in 5 U.S.C. § 22.

28. See *Investigation of Charges that Proposed Security Regulation under Executive Order 9835 Will Limit Free Speech and a Free Press*, Hearings before a Subcommittee of the House Committee on Expenditures in the Executive Departments, 80th Cong., 1st Sess. (1947), pp. 4, 13; and see also Fritz Morstein Marx, "Effects of International Tension on Liberty under Law," 48 *Columbia Law Review* 555, 560 *et seq.* (1948).

29. "Document," as that term is used by the Army, includes, among other

things, "printed, mimeographed, typed, photostated, and written matter of all kinds; . . . correspondence and plans relating to research and development projects, and all other similar matter." 10 CFR (Cum. Supp.) 5.1(b). The Navy's regulations are similarly inclusive, though they make no specific reference to research and development projects. Navy Regs. (Rev. 1944), art. 76(1)(a).

30. The quoted expression in favor of moderate classification is found in Department of the Army, 10 CFR 5.1(b).

31. *Administration for Research,* note 13 above, p. 36. See also *National Security and Our Individual Freedom, A Statement on National Policy* by the Research and Policy Committee of the Committee for Economic Development (December 1949), p. 23: "More important than the letter of a regulation is the spirit in which it is administered. At present, there is one-sided emphasis upon the importance of secrecy in the indoctrination of officers both military and non-military. A government official is rarely commended for disclosure. He may, however, be reprimanded or otherwise disciplined for 'under-classification,' that is, for failure to make material confidential or secret."

32. Columbia's difficulty in its copper chloride study is described in W. A. Noyes, Jr. (ed.), *Chemistry,* Science in World War II (Little, Brown & Co., 1948), p. 433.

33. The Navy's regulations furnish the following summary of disclosure policy, illustrative of the effective narrowing of the range of transmissibility of classified information:

"Information as to the existence, nature or whereabouts of 'secret' matter shall, except as specifically authorized by the Chief of Naval Operations, be disclosed only to those persons in the Government service whose official duties require such knowledge. 'Confidential' matter may be disclosed to persons in the Government service who must be informed, and to other persons therein when, under special circumstances, such disclosure is to the interest of the Navy. 'Restricted' matter may be disclosed to persons of discretion in the Government service when it appears to be in the public interest.

"Information as to the existence, nature, or whereabouts of 'secret' matter, shall, except as authorized by the Chief of Naval Operations, be disclosed only to persons not in the Government service who under conditions of absolute necessity must be informed. . . ." U.S. Naval Regs. 75½(4)(b) and (c).

The Army regulations provide more broadly that classified information may be discussed with governmental personnel and private individuals who have a legitimate interest in it, though there is no suggestion of what constitutes a legitimate interest or who is to determine whether it exists. 10 CFR (1944) 105.2(b).

34. The Army Contract provision appears in CFR (1947 Supp.) Title 10, 805.401–2; the Navy clause, having similar purpose, appears in Naval Procurement Directives, March 16, 1944, Enclosure C, 11261 C.

35. Army and Navy regulations bearing upon the declassification process may be found in 10 CFR 5.1(b)(7); U.S. Naval Regs. 75½(2)(b) and 75½(2)(c).

36. The persistence of no longer defensible classifications is discussed by Stewart, note 20 above, p. 252.

37. The relevant orders about publishing information captured from the enemy are Executive Orders No. 9568, June 8, 1945, and No. 9604, August 25, 1945, CFR Supp. 1945, Vol. 3, pp. 78, 108.

38. At the end of June 1950, however, the Air Force through its School of Aviation Medicine released two volumes, 1,300 pages, devoted to *German Aviation Medicine, World War II.* These volumes reflected German researches during the years 1939–1945, and described equipment and data on "researches of general interest in physiology, biophysics, psychology and pathology." *New York Times,* June 27, 1950, p. 53, col. 6.

39. The fate of the "Summary Technical Reports" is discussed by Stewart, note 20 above, p. 291. On May 22, 1950, a portion of one volume of the "Summary Technical Reports" was declassified and was then published by the AEC as a *Handbook on Aerosols* because the wartime research on the behavior of dusts, fumes, and mists had an immediate bearing on preventing atmospheric contamination by radioactivity. AEC Release No. 285.

Chapter II

1. Vannevar Bush, *Modern Arms and Free Men* (published by Simon & Schuster; copyright, 1949, by The Trustees of Vannevar Bush Trust), p. 101.

2. A description of wartime research in connection with the nitrogen mustards appears in W. A. Noyes, Jr. (ed.), *Chemistry,* Science in World War II (Little, Brown & Co., 1948), pp. 166–168, 243, 247, 250, 251, 256–258.

3. The BAL story is pieced together from the following sources: R. A. Peters, L. A. Stocken, and R. H. S. Thompson, "British Anti-Lewisite (BAL)," 156 *Nature* 616 (1945); H. Eagle and H. J. Magnuson, "Systematic Treatment of 227 Cases of Arsenic Poisoning," 30 *American Journal of Syphilis, Gonorrhea, and Venereal Diseases* 420 (1946); W. T. Longcope and J. A. Leutscher, Jr., "Treatment of Acute Mercury Poisoning by BAL," 25 *Journal of Clinical Investigation* 557 (1946); C. Ragan and R. H. Boots, "Treatment of Gold Dermatitides with BAL," 133 *American Medical Association Journal* 752 (1947).

4. Dr. Compton's remarks on the retarding effects of secrecy appear in *Hearings before Senate Committee on Military Affairs on S. 1297,* 79th Cong., 1st Sess. (1945), p. 625.

5. The AEC commented on its compartmentalization policy in its *Fifth Semiannual Report to Congress* (published by the Government Printing Office as *Atomic Energy Development, 1947–1948*), pp. 83, 84.

6. Compartmentalization in radar research is discussed in L. N. Ridenour, "Secrecy in Science," 1 *Bulletin of the Atomic Scientists,* No. 6, p. 3 (1946); also in *Hearings before Senate Special Committee on Atomic Energy,* 79th Cong., 1st Sess. (1945), pp. 538, 539, 542. And compare E. U. Condon, "Science

and Security," 107 *Science* 659, 662 (1948): "With the microwave field at the Radiation Laboratory in Cambridge, Massachusetts, there was no compartmentalization whatever . . . More than that, there were frequent secret conferences on special topics, attended by hundreds of staff members. People in all parts of the subject went to a great deal of trouble to keep those in other parts fully informed. I believe that a great deal was gained by this lack of compartmentalization in the field of microwave radar." Dr. Condon adds the observation that in the atomic bomb project, compartmentalization prevented the acquisition of data that were badly needed by project workers, but that the British scientists (who were not hampered by compartmentalization rules) were able to supply some of the desired information; he expresses the belief that "we would have had a much harder time with the atomic bomb project had our British friends not short-circuited compartmentalization for us."

7. Naval fire-control difficulties are discussed in Joseph C. Boyce (ed.), *New Weapons for Air Warfare*, Science in World War II (Little, Brown & Co., 1948), p. 95.

8. Dr. Manley's comments on compartmentalization appear in his article, "The Los Alamos Scientific Laboratory," 5 *Bulletin of the Atomic Scientists* 101, 105 (1949).

T. R. Hogness, director of the Institute of Radiobiology and Biophysics, University of Chicago, in an address on "Security, Secrecy, and the Atom Bomb," delivered before the American Veterans Committee on November 25, 1949, attributes the Los Alamos policy to its former director, J. R. Oppenheimer. Oppenheimer "argued that the design of a bomb was too great a responsibility for just a few men. He needed the ideas of many, and many of the best ideas came from unexpected sources. Had the hierarchic attitude been adopted at Los Alamos, we might not have had the bomb."

9. Mervin J. Kelly, executive vice-president of the Bell Laboratories, served as an AEC consultant in the early summer of 1949. He later testified before a Congressional committee that "within the remainder of the atomic energy activities area, by that I mean Oak Ridge, Argonne, Hanford, I found good liaison and good cross-connections of knowledge between the programs. Actually, the week after I left there was an internal scientific meeting at Los Alamos of the scientists from these different laboratories, all of them being cleared, and, therefore, they could talk about the basic physics that was fundamental to this job. They were having a meeting much like the physical society meetings, except on classified material, and the contacts on matters of business on the technical things that flow between these organizations were in very good standing and being well done." *Investigation into the United States Atomic Energy Project*, Hearing before the Joint Committee on Atomic Energy, 81st Cong., 1st Sess. (July 7, 1949), p. 812.

It is only proper to add, nevertheless, that the scientists at other installations have steadily maintained, contrary to Dr. Kelly's impression, that they do not receive adequate information concerning the work done at Los Alamos. The meeting to which Dr. Kelly refers is apparently one of the semiannual "information meetings" at which scientists from the various installations discuss restricted data, not including, however, data that have a bearing on weapons, a rather large exclusion.

10. The comment on the educational values of a research failure comes from E. H. Land, president and research director of the Polaroid Corporation, in *The Future of Industrial Research* (Standard Oil Co., 1945), p. 84.

11. Sir Alexander Fleming commented on unsuccessful research in his Dedication Address, Oklahoma Medical Research Foundation, July 3, 1949.

12. *Report to the U.S. Atomic Energy Commission* by the Industrial Advisory Group, in 71 *Mechanical Engineering* 205, 208–209 (1949); also in 5 *Bulletin of the Atomic Scientists* 51, 54 (1949).

Subsequently, and perhaps influenced by the above-cited report, the AEC has somewhat relaxed its restraints upon industrial information. Thus on October 21, 1949, at the sessions of the National Metal Congress and Exposition in Cleveland a number of technical papers were read by AEC researchers, giving to the assembled manufacturing experts a considerable amount of previously unavailable research information on alloys and metals, including uranium, thorium, and beryllium.

13. The comment on scientific interrelations in reactor design is by F. H. Spedding, "Chemical Aspects of the Atomic Energy Problem," 5 *Bulletin of the Atomic Scientists* 48, 49 (1949).

14. *The Report of the Committee on the National Security Organization*, Commission on Organization of the Executive Branch of the Government (1948), III, 151, emphasizes that while it is important to save money by avoiding unnecessary duplication of research, the more important thing to the nation is the risk that we have not adequately developed "skill in making and utilizing scientific advances."

15. Dr. Lawrence S. Kubie's comments on German research habits may be found in *Hearings before Senate Committee on Military Affairs on S. 1297*, 79th Cong., 1st Sess. (1945), p. 618.

16. The quoted conclusion about the costs of compartmentalization is from John E. Burchard (ed.), *Rockets, Guns and Targets*, Science in World War II (Little, Brown & Co., 1948), p. 322.

17. Irvin Stewart, *Organizing Scientific Research for War: The Administrative History of the Office of Scientific Research and Development* (Little, Brown & Co., 1948), p. 28.

18. *Ibid.*, p. 29. For a similar explanation along with suggestive discussion of the deadening consequences of compartmentalization, see Joseph C. Boyce, note 7 above, pp. 98–100.

19. The Japanese military's attitude toward scientists is discussed in James Phinney Baxter III, *Scientists against Time* (Little, Brown & Co., 1946), p. 10: "Both services distrusted the civilian scientists, especially if they had been educated in America, England, or even Germany. They consequently refused to

give them sufficient information, and hampered research by security regulations pushed to the limits of fantasy. . . ."

The head of the Board of Technology of the Japanese War Cabinet is quoted as having told Karl T. Compton, former chairman of the Research and Development Board of the National Military Establishment, that "There was no cooperation between the army and the navy. A general would rather lose the war than shake hands with an admiral. And as for our scientists, we were treated by the military almost as if we were foreigners." Address by Dr. Compton, "Science and National Strength: Some Lessons from World War II," delivered at the Aeroballistics Facility, Naval Ordnance Laboratory, June 27, 1949.

20. Boyce, note 7 above, p. 275.

21. For a suggestion that the AEC publicists "sometimes wrap newsworthy activities in the same fog of misunderstanding they are supposedly on hand to dispel," see Layton Lewis, "The Fifth Report: A Press View," 5 Bulletin of the Atomic Scientists 93 (1949).

22. The episode of the concrete structures and the bombs is discussed in Burchard, note 16 above, p. 318.

T. R. Hogness, director of the Institute of Radiobiology and Biophysics of the University of Chicago, asserted in an address before the American Veterans Committee on November 25, 1949: "Secrecy can also be used as a cloak to cover inefficiency. Between World Wars I and II some of the branches of the Army were operated very inefficiently. It is sufficient to remind you that effective tanks were developed by our country only when the last war was well along. But who knew about this inefficiency other than the departments involved?"

23. The Industrial Advisory Group was appointed in October 1947; its report is dated December 15, 1948. The text of the report is printed in full in 71 Mechanical Engineering 205 (1949) and in part in 5 Bulletin of the Atomic Scientists 51 (1949).

24. The survey of scientists' opinions was reported at full length in Administration for Research (Vol. III of Science and Public Policy), Report of the President's Scientific Research Board (October 4, 1947), Appendix III, pp. 205–252.

25. See, e.g., Taking a Chance, AEC Security Pamphlet No. 2 (1948), a widely circulated leaflet that grimly tells the story of a man who, after working briefly in one of the secret laboratories, had written a monograph which might have "revealed to an inquiring mind secrets that might be of value to another nation." "Possibly," the pamphleteer adds, "he reasoned that because the authorities at Oak Ridge had removed all secret data from his notebooks he was completely free to use in any way everything that remained. He forgot, of course, that they had not excised his memory and that everything he wrote might be flavored by that memory."

Samuel K. Allison, wartime director of the "Metallurgical Laboratory" of the Manhattan Project, has expressed the belief that "the existence of an inner core of secret facts vitiates whole areas of scientific inquiry and technological development extending far from the actually classified data. No one can remember from day to day just what is classified, and to be safe, avoids discussing whole fields of research and technology." "The State of Physics; or the Perils of Being Important," 6 *Bulletin of the Atomic Scientists* 2, 4 (1950).

The AEC itself has reinforced the disinclination of scientists to discuss wholly nonsecret matters. On March 14, 1950, it peremptorily directed all its contractors (including universities) to tell their employees to avoid discussion of all technical information bearing on thermonuclear weapons (hydrogen bombs), whether classified or not. Several days later this direction was somewhat "toned down" and became merely a request rather than an abrupt command. See a review of this matter in *New York Times*, March 30, 1950, p. 1, col. 6. More recently Commissioner H. D. Smyth reportedly expressed the opinion that it "makes a great deal of difference who is giving out information" and that men who have had access to classified data should "realize that they themselves cannot always make a sound judgment on the significance of what they have written, however well acquainted they are with certain phases of the project." *New York Times*, April 29, 1950, p. 17, col. 8.

26. The lament about the lack of allure in a military-scientific career appears in *Scientists in Uniform, World War II*, A Report to the Deputy Director for Research and Development, Logistics Division, General Staff, U.S. Army (1948), p. 31.

27. A recent poll of scientists, conducted by the National Opinion Research Center, University of Denver, revealed that "among scientists employed by the Federal Government, only 37 percent felt that the greatest career satisfaction could be obtained in the Government; only 5 percent and 1 percent respectively of industrial and university scientists agreed with them. Of all groups combined, only 11 percent preferred a Government career in terms of satisfaction, while 31 percent preferred industry and 48 percent the university environment. The remaining 10 percent preferred consulting work or some other activity." *Administration for Research* (Vol. III of *Science and Public Policy*), A Report to the President by John R. Steelman, Chairman, the President's Scientific Research Board (October 4, 1947), Appendix III, p. 205.

28. Dr. Compton is quoted by John E. Pfeiffer in "Top Man in American Science," *N.Y. Times Magazine*, Oct. 17, 1948, p. 68.

29. As an example of plain silliness the following United Press dispatch from Washington, dated September 6 and appearing in the *New York Times* of September 7, 1948, p. 20, col. 6, warrants preservation:

"The House Committee on Un-American Activities is trying to find out why a group of scientists has chosen a part of Africa, rich in uranium, to set up a $9,000,000 astronomical laboratory.

"Representative John McDowell, Republican, of Pennsylvania, a committee member, stated that an investigator for the House Group was looking into a venture involving American, Belgian, French and Dutch scientific interests.

" 'We are not undertaking this check-up as an attack on science, but in these days it is essential to learn such things as the source of the financing and who is behind the whole business,' Mr. McDowell said."

The next day the Belgian Colonial Ministry rather testily announced that two Belgian astronomers, Raymond Courtrez and Lucien Boss, were in the Belgian Congo to find an emplacement for an observatory, sponsored by the official institute for scientific research in Central Africa. "These activities are purely scientific in character and have nothing to do with the Belgian Congo uranium. They do not justify any investigation by the Un-American Activities Committee of the United States Congress."

The House Committee's interest stemmed from the fact that the internationally famed Harvard astronomer, Professor Harlow Shapley, was among the directors of the research group; he is not, however, among those who are dearly beloved by the House Committee. There is no way of telling whether the Committee's inability to differentiate between, on the one hand, the fissionable isotopes of uranium and, on the other, large natural deposits of wholly nonexplosive uranium ore is attributable to deliberate distortion or merely to dismal lack of knowledge.

30. Discussion of the unattractiveness of the government's research program appears in *Administration for Research* (Vol. III of *Science and Public Policy*), A Report to the President by John R. Steelman, Chairman, the President's Scientific Research Board (October 4, 1947), p. 162.

31. *Ibid.,* p. 163. The Steelman report notes in passing that "some contributions of civilian scientists in the Office of Scientific Research and Development were withheld from the public during the war for security reasons. They were, however, revealed at the end of the war with a lion's share of the credit to the military establishments rather than to the scientists actually responsible for the work." Naturally enough the scientists did not join in the applause. *Ibid.,* p. 165.

32. The AEC's comment on the denial of opportunity to publish researches appears in its *Fifth Semiannual Report to Congress (Atomic Energy Development 1947–1948),* p. 107 (1948).

33. I. I. Rabi, "Publication Is the Chief Responsibility," 4 *Bulletin of the Atomic Scientists* 73 (1948).

34. On the subject of withholding opportunity to acquire prestige, compare the following comment in the Steelman report, cited in note 30 above, p. 164: "A major factor in the professional recognition of a scientist is his attendance and presentation of papers at meetings of professional scientific societies. Both the scientist and the Government gain prestige and recognition by adequate representation at such meetings. Despite this fact, attendance at meetings is limited by lack of travel funds in most scientific units of the Government . . . It appears to be penny-wise and pound-foolish to pay a man several thousand dollars a year for his special scientific competence and then deny him the means to maintain that competence at a high level for the Government's benefit.

"Not only is this policy uneconomic as it applies to scientists, but it fails to recognize that progress in science depends in the final analysis upon intellectual stimulation. As J. R. Oppenheimer stated before a congressional committee in October 1945, '* * * the gossip of scientists who get together is the lifeblood of physics, and I think it must be in all other branches of science * * *.'"

For recognition of the publication problem as it affects Los Alamos, see testimony of Dr. Norris E. Bradbury, director of the laboratory there, in *Investigation into the United States Atomic Energy Project*, Hearings before the Joint Committee on Atomic Energy, 81st Cong., 1st Sess. (July 7, 1949), pp. 820–822.

The Joint Congressional Committee on Atomic Energy recently reported that "The adverse effect of secrecy upon scientific morale is being reduced through periodic seminars and conferences attended exclusively by persons who possess security clearance. Dr. Bradbury depicted these sessions as a vehicle whereby Commission experts not only exchange ideas and stimulate one another's thinking but also gain recognition, within the limits of the cleared group, for accomplishments which once might have attracted the applause of scientists generally. Circulation of technical papers among cleared personnel produces the same result. An ambitious young physicist is, therefore, less likely to reject atomic energy employment for fear that secrecy would prevent him from building a reputation." *Investigation into the United States Atomic Energy Commission*, 81st Cong., 1st Sess., Senate Rep. No. 1169 (1949), p. 36.

35. The difficulty of introducing students into the research training program was discussed by Robert M. Boarts, professor of chemical engineering, University of Tennessee, in an address entitled "Nucleonics and the Graduate Program in Chemical Engineering," delivered before the American Society for Engineering Education, June 15, 1948. Recently the AEC has established a reactor development training school at Oak Ridge; the student body, numbering 90, will be made up of industrial engineers, government employees, and recent college graduates. The school was opened to "meet the need for that rather unique combination of engineer and physicist so necessary to provide talent for the rapidly growing reactor field." AEC Commissioner Gordon Dean, in an address entitled "Atomic Energy in War and Peace," delivered before the American Medical Association, June 26, 1950.

36. Testimony of Enrico Fermi before the Joint Committee on Atomic Energy, *Investigation into the United States Atomic Energy Project*, Hearing, 81st Cong., 1st Sess. (1949), p. 871.

The transcript of an AEC press conference on March 29, 1950, shows at page 8 the following comment by Commissioner Smyth: ". . . in order to prepare for the technological development of 5 or 10 or 15 years from now you have to have men trained in universities. And it is very difficult to train men when you have secret projects going on. I might illustrate this by telling you that during the war we had a course at Princeton in nuclear physics. We had to look around to find somebody on our staff who had no connection with the Manhattan Project, because no one who was working for the Manhattan Project would dare to try to separate in his mind what he could say and what he couldn't say. The result was that—with all due respect to the man we got to give the course on nuclear physics—he wasn't an authority on nuclear physics, and those men didn't get very good training."

37. Dr. Morse stated his conclusions about education in nuclear engineering in an address before the 1948 New York Herald–Tribune Forum, reported in the *N.Y. Herald-Tribune*, October 24, 1948, sec. X, p. 56, col. 5.

38. Dr. Bacher discussed the need for trained personnel in his testimony before the Joint Committee on Atomic Energy, *Investigation into the United States Atomic Energy Project,* Hearing, 81st Cong., 1st Sess. (July 6, 1949), p. 783.

Chapter III

1. The comment on the need of having new secrets is by E. U. Condon, who added, "The price we have to pay in order to grow in knowledge is some giving up of present knowledge in order that we may continue to grow." "Science and Security," 107 *Science* 659, 660 (1948).

2. Sir Alexander Fleming's remarks on penicillin manufacture occurred during his Dedication Address, Oklahoma Medical Research Foundation, July 3, 1949.

3. The distinction between principle and process was brought out by F. H. Spedding, director of the Atomic Research Institute at Iowa State College, "Chemical Aspects of the Atomic Energy Problem," 5 *Bulletin of the Atomic Scientists* 48, 49 (1949).

4. The so-called "Merck Report" was issued by the War Department on January 3, 1946. The full text appears under the heading, "Official Report on Biological Warfare," in 2 *Bulletin of the Atomic Scientists,* Nos. 7–8, p. 16 (1946). The report was subsequently withdrawn from circulation by the War Department. For three years no statements concerning BW emanated from the military. The next release on the subject was a brief and general statement by Secretary of Defense James Forrestal, March 12, 1949, intended to counteract exaggerated impressions concerning the potency and state of development of biological warfare. The Forrestal statement is printed in 5 *Bulletin of the Atomic Scientists* 104 (1949). Then, for more than a year, the subject lapsed back into the silences in which the Army has habitually enveloped it. Secretary of Defense Louis Johnson next mentioned BW in his semiannual report to the President dated March 31, 1950, at pp. 69–71. He remarked that "complete" and "detailed" studies had been made concerning a number of disease agents infectious for man, domestic animals, and crop plants, but that it would be unwise from a security viewpoint to publish these studies.

5. The list of BW research accomplishments is taken from the official Merck Report, 2 *Bulletin of the Atomic Scientists,* p. 18.

6. For fuller treatment of BW researches and their beneficial possibilities, see Theodor Rosebury, *Peace or Pestilence* (Whittlesey House, 1949), pp. 186 ff.

7. Some of the BW work, it may be noted in passing, illustrates the efficiency which flows from noncompartmentalization of scientific effort. An introductory note to one of the series of reports, after remarking the varied personnel that shared in the researches, asserts: "The highly successful outcome of the work in developing protective measures against rinderpest, one of the most devastating diseases of cattle, including improved methods of vaccine production plus fundamental observations significant to virus-disease research, constitute an outstanding contribution to veterinary science and another shining example of what can be accomplished through collaboration of scientists from several fields." R. E. Shope *et al.*, "Papers on Rinderpest Virus," 7 *American Journal of Veterinary Research* 133 (1946).

8. For discussion of Alloy X and of other developments which are of potential industrial interest, see John E. Burchard (ed.), *Rockets, Guns and Targets,* Science in World War II (Little, Brown & Co., 1948), pp. 394, 422–423.

9. The development of sabotage devices and of security restrictions on them is discussed by W. C. Lothrop, "History of Division 19: Weapons for Sabotage," in W. A. Noyes, Jr. (ed.), *Chemistry,* Science in World War II (Little, Brown & Co., 1948), pp. 434, 437.

10. Robert F. Bacher, former AEC Commissioner, speaking on the subject, "Our Progress in Atomic Energy," at Los Angeles Town Hall, October 3, 1949, said: "A good many of the developments in atomic energy have been shrouded in a veil of secrecy. Information about the design and production of weapons and the production of fissionable material has been very closely held. But the veil of secrecy has a tendency to spread like a fog and cover all sorts of other subjects as well. No one wants to be responsible for making information generally available which someone might claim should remain secret. As a result, many developments are kept secret which might have led to major advances elsewhere in American industry."

11. An address by then Commissioner Robert F. Bacher before the American Academy of Arts and Sciences on February 9, 1949, contained extensive discussion of the reactor program of the AEC. R. F. Bacher, "The Development of Nuclear Reactors," 5 *Bulletin of the Atomic Scientists* 80 (1949). Without detailing matters of design, he described the types, purposes, and limitations of the nuclear reactors then in existence or in contemplation. His candid exposition reflected a trend that was apparent also in the Commission's *Fifth Semiannual Report.* Further, the AEC took the initiative in discussions with the Air Forces, the Bureau of Aeronautics, and the National Advisory Committee on Aviation looking toward release of basic information in the so-called Lexington Report on the feasibility of developing a reactor to propel an aircraft. See testimony of Carroll L. Wilson, AEC General Manager, in *Hearing before the Joint Committee on Atomic Energy,* 81st Cong., 1st Sess. (February 2, 1949), pp. 25–26.

12. Hearing cited in note 11 above, pp. 14–17.

13. Karl T. Compton, then the chairman of the Research and Development

Board of the National Military Establishment, has summarized the conflict between secrecy and science in the following terms:

"Unfortunately, secrecy and progress are mutually incompatible. This is always true of science, whether for military purposes or otherwise. Science flourishes and scientists make progress in an atmosphere of free inquiry and free interchange of ideas, with the continual mutual stimulation of active minds working in the same or related fields. Any imposition of secrecy in science is like application of a brake to progress. . . . It is much easier for the average citizen to understand secrecy than it is for him to understand the conditions necessary for scientific progress. I am sure that the pendulum has recently swung so far in the direction of concern over secrecy regarding even little details and unimportant people that our real security is suffering. It is suffering from the slowing up of progress because attention is being diverted from the really big things which need to be done." Dedication Address at the Aeroballistics Facility, Naval Ordnance Laboratory, White Oak, Maryland, June 27, 1949.

Chapter IV

1. Figures on the AEC's devotion of time to security problems are given in the report by the Joint Committee on Atomic Energy, *Investigation into the United States Atomic Energy Commission,* 81st Cong., 1st Sess., Senate Rep. No. 1169 (1949), p. 85.

2. General Donovan commented on the need of taking some calculated security risks in a press interview reported in the *New York Times,* August 31, 1948, p. 3, col. 8.

3. Figures furnished by the AEC in *Hearings before the Joint Congressional Committee on Atomic Energy,* 81st Cong., 1st Sess. (February 2, 1949), p. 30, show that the Personnel Security Review Board met only once between July 1, 1947, and December 31, 1948, at a total travel cost of $151.94. I am unable to account for the discrepancy between the Board's minutes and the Commission's records, but believe that the former are more reliable in this instance.

4. The AEC General Manager's instructions concerning derogatory information are contained in Bulletin GM-80, dated March 30, 1948. Though the document bears no indication that it is classified, it has apparently never been published and a number of officials have declined to discuss its contents as though they were matters of high policy. The text of the Bulletin, which was subsequently obtained, does not warrant the secretiveness with which it has at times been surrounded.

5. The full text of the "Criteria" appears in Appendix B, pp. 238–244 of this volume. It was officially published in 14 *Federal Register,* No. 3, p. 42 (1949).

6. The Joint Congressional Committee on Atomic Energy reported on October 13, 1949, that, of the 150,000 investigations which the FBI had by then

completed, only 2,125, or one in every seventy, brought forth any data that required special attention, and these were facts "usually involving character alone." *Senate Report No. 1169,* 81st Cong., 1st Sess. (1949), p. 66.

7. The summary power to remove employees of the military departments was conferred by Public Law 808, 77th Cong., 2nd Sess., 56 Stat. 1053, § 3.

8. The comment on the importance of scientific resources is by A. C. Mc-Auliffe, Major General, GSC, in a foreword to *Scientists in Uniform, World War II,* Report to Deputy Director for Research and Development, Logistics Division, General Staff, U.S. Army (1948).

9. Announcement concerning the composition, procedures, and decisional standards of the Industrial Employment Review Board was made by Secretary of Defense Johnson in a press release (Rel. No. 544–49) on December 5, 1949.

10. The Eisenhower directive concerning "suspension of subversives" is printed in War Department Pamphlet 32-4, December 10, 1946.

11. The criteria of judgment concerning personnel security in 1948 were published as Army Mem. No. 380-5-10, April 2, 1948, p. 9.

12. The Criteria, the full text of which can not be found in the Federal Register or the Code of Federal Regulations, state that access shall be denied "if, on all the evidence and information available to the Board, reasonable grounds exist for belief that the individual: . . . 6. Is or recently has been a member of, or affiliated or sympathetically associated with, any foreign or domestic organization, association, movement, group, or combination of persons (a) which is, or which has been designated by the Attorney General as being, totalitarian, fascist, communist or subversive, (b) which has adopted, or which has been designated by the Attorney General as having adopted, a policy of advocating or approving the commission of acts of force or violence to deny other persons their rights under the Constitution of the United States, or (c) which seeks, or which has been designated by the Attorney General as seeking, to alter the form of the government of the United States by unconstitutional means; provided, that access may be granted, notwithstanding such membership, affiliation or association, if it is demonstrated, by more than a mere denial, that the security interests of the United States will not thereby be jeopardized."

13. The Secretary of Defense and the military secretaries apparently agree with this observation, for there is only one limitation upon the appointive power that has been lodged in the secretaries of the Departments of the Army, Navy, and Air Force, namely: "No person who has served with an investigative agency of any of the Departments within one year preceding his appointment shall be eligible for appointment as a member or alternate member of the Board . . ."

14. *Duncan* v. *Kahanamoku,* 327 U.S. 304, 324 (1946).

15. Discussion of recent experiences with military tribunals of which the United States must be ashamed may be found in A. Frank Reel, *The Case of General Yamashita* (University of Chicago Press, 1949); and see the description of the military judicial system in Hawaii by Attorney General Garner Anthony

in a report to Governor Ingram M. Stainback, December 1, 1942, quoted by John P. Frank in "Ex parte Milligan v. The Five Companies: Martial Law in Hawaii," 44 *Columbia Law Review* 639, 652 (1944).

16. In a recent statement on national policy the Research and Policy Committee of the Committee for Economic Development stressed the conviction that civilian supremacy is essential to freedom and that "without effective civilian control there is danger that security policy will be made more and more by the military alone and in terms of the individual problems of military defense for which they are responsible rather than in the larger terms of security and freedom." C.E.D., *National Security and Our Individual Freedom* (December 14, 1949), pp. 6, 11.

Chapter V

1. *Science and Foreign Relations,* Dept. of State Publication No. 3860, May 1950, at p. 80. Chapter VI of that document, a report by Dr. Lloyd V. Berkner reviewed by the Department's Steering Committee and unanimously approved by the Advisory Committee of the National Academy of Sciences, is devoted to "Control Over the International Flow of Scientific Information— Persons and Material." It contains other examples of the extension of security practices into wholly unclassified areas of activity.

2. *Hearings Regarding Communist Infiltration of Radiation Laboratory and Atomic Bomb Project at the University of California, Berkeley, Calif.,* House Committee on Un-American Activities, 81st Cong., 1st Sess. (1949), pp. 280 ff.

3. An excellent analysis of the press treatment of the Committee's charge that Dr. Edward U. Condon, director of the National Bureau of Standards, is "one of the weakest links in our atomic security," has been completed by the Columbia University Bureau of Applied Social Research, under the direction of Prof. Robert K. Merton. It is fully reported by J. T. Klapper and C. Y. Glock in "Trial by Newspaper," 180 *Scientific American,* No. 2, p. 16 (1949).

4. Brandeis, J., dissenting with Holmes, Butler, and Stone, JJ., in *Olmstead v. United States,* 277 U.S. 438, 478 (1928).

5. The Joint Congressional Committee on Atomic Energy recently reported that 874 persons had withdrawn applications for clearance because, before action in their cases had been completed, they decided to work elsewhere. *Senate Rep. No. 1169,* 81st Cong., 1st Sess. (1949), p. 66. It must be stressed, however, that not all of these 874 persons were scientists.

6. S. T. Pike, "Witch-Hunting Then and Now," 180 *Atlantic Monthly* 93, 94 (1947). And compare C. E. Merriam, "Some Aspects of Loyalty," 8 *Public Administration Review* 81, 84 (1948): "The basis of modern scientific and technological progress which is the key to our civilization is not found in complete conformity and docility, but in critical attitudes leading to invention and advance in public as well as in private business. We must be on the alert for unorthodox,

original, creative minds, capable of discovering new relations and better ways of doing things in peace as well as in war."

Chapter VI

1. The pertinent references to laws governing misconduct of the types described in Executive Order No. 9835 will be found in a comprehensive and penetrating article by T. I. Emerson and D. M. Helfeld, "Loyalty among Government Employees," 58 *Yale Law Journal* 1, 27 ff. (1948).

2. The statutory bar against employment of Communists in the federal service is found in the Hatch Act, § 9A, 53 Stat. 1147, 1148 (1939); *H.R. Rep. No. 616*, 80th Cong., 1st Sess. (1947), p. 4.

3. The Supreme Court's views on the meaning of "affiliation" are expressed in *Bridges* v. *Wixon*, 326 U.S. 135, 143–144 (1945).

4. The President's statement about the significance of membership in an organization is set forth in the *New York Times,* November 15, 1947, p. 2, col. 3; its substance is repeated in the official "Regulations for the Operations of the Loyalty Review Board," 13 Fed. Reg. 253, 254 (1948). In this respect the Loyalty Order is considerably less drastic than the statute that created the Economic Cooperation Administration. Section 110-c of the Foreign Assistance Act of 1948, 62 Stat. 142, 22 U.S.C. § 1508-c, provides that no one may be employed until after investigation by the FBI and certification by the Administrator that the individual "is loyal to the United States, its Constitution, and form of government, *and is not now and has never been a member of any organization advocating contrary views."* This restriction, embodying what might be called the doctrine of perpetual guilt, might well be deemed an unconstitutional bill of attainder within the holdings of the Supreme Court in *Ex parte Garland*, 4 Wall. 333 (1867) and *Cummings* v. *Missouri*, 4 Wall. 277 (1867), in the latter of which Mr. Justice Field said: "The theory upon which our political institutions rest is, that all men have certain inalienable rights—that among these are life, liberty, and the pursuit of happiness; and that in the pursuit of happiness all avocations, all honors, all positions, are alike open to every one, and that in the protection of these rights all are equal before the law. Any deprivation or suspension of any of these rights for past conduct is punishment, and can be in no otherwise defined."

5. Chairman Richardson's comments on joining organizations are reported in the *New York Times,* Dec. 28, 1947, p. 28, col. 6.

6. Senator Taft's identification of the Democratic Party with totalitarianism is reported in the *New York Times,* Feb. 4, 1949, p. 13, col. 4.

7. For comment upon the days when veterans hospitals were deemed communistic see H. N. Rosenfield, "Experts Are Never Right," *Antioch Review,* Spring 1949, pp. 3, 6.

8. President Truman's denunciation of the real estate groups as subversive is recorded in the *New York Times,* July 1, 1947, p. 20, col. 8.

9. Mr. Hoover's opinion about totalitarianism is reflected in the *New York Times*, June 23, 1948, p. 8, col. 5.

10. See, e.g., "Designation of Organization as Subversive by Attorney General: A Cause of Action," 48 *Columbia Law Review* 1050 (1948). The first appellate decision handed down was *Joint Anti-Fascist Refugee Committee* v. *Clark*, 177F. (2d) 79 (District of Columbia Court of Appeals, 1949); Judges Proctor and Bennett Clark concluded that the black list could not be attacked by an organization that was placed on it; Judge Edgerton dissented. The Supreme Court has agreed to review this decision when it convenes in the autumn of 1950; certiorari was granted, 339 U.S. 910 (1950). *National Council of American Soviet Friendship, Inc.* v. *McGrath*, involving the same issues as the *Joint Anti-Fascist Refugee Committee* case, was decided by the Court of Appeals without opinion on October 25, 1949; and certiorari has been granted in that case as well, 70 Sup. Ct. 978. *International Workers Order* v. *McGrath*, decided by the Court of Appeals on March 22, 1950, has not yet been officially reported, but may be found in Pike-Fischer *Administrative Law* 52a. 21–36. A petition for certiorari has been filed by the IWO in that case, but had not been acted on by the Supreme Court prior to its adjournment in the summer of 1950.

11. The Attorney General has, however, recently modified the listing of the Societá Dante Alighieri as a fascist group. According to the Italian Embassy, the Department of Justice has notified the society that its inclusion among the fascist organizations on the black list "does not apply" to the group as it is now constituted nor "to any of its activities since its re-establishment at the end of the second World War." See *New York Times*, November 13, 1949, p. 9, col. 5.

12. The President's characterization of the loyalty probe is reported in the *New York Times*, November 15, 1947, p. 2, col. 2.

13. H. S. Commager, "Who Is Loyal to America?", 195 *Harper's Magazine* 193, 198 (1947).

14. The Loyalty Review Board's differentiation between permissible advocacy and impermissible allegiance is embodied in a statement of Chairman Richardson, *New York Times*, December 28, 1947, p. 28, col. 6.

15. The long lists of the House Committee may be found in its document entitled *Citations by Official Government Agencies of Organizations and Publications Found to Be Communist or Communist Front* (1948), p. 1.

16. The views of the California Committee, under the chairmanship of Senator Tenney, concerning the American Civil Liberties Union, appear in the *Fourth Report of the Senate Fact-finding Committee on Un-American Activities* (1948), pp. 107 ff.

17. See the scholarly and exciting study of Prof. Marion L. Starkey, *The Devil in Massachusetts: A Modern Inquiry into the Salem Witch Trials* (Alfred A. Knopf, 1949).

18. The Loyalty Review Board's Directive II started the ball rolling in the direction of issuing charges and holding hearings unless the case was so alto-

gether plain as to be considered "clearly favorable." This was reinforced on September 29, 1949, by Memorandum No. 48, which again stressed that notices of charges should be issued to employees "in cases in which a finding clearly favorable to the individual is not clearly warranted." The loyalty boards were instructed to cease trying to find out what sort of finding was warranted by any means short of a hearing.

General Donovan has criticized this approach, saying: "Under the existing system, many cases of no substance reach the Loyalty Board which must then take on the first responsible job of eliminating unwarranted complaints. Doing this at an earlier stage would alleviate the burden of work placed on the Loyalty Boards and relieve the employee from the harassment of a protracted Loyalty hearing." W. J. Donovan and M. G. Jones, "Program for a Democratic Counter Attack to Communist Penetration of Government Service," 58 *Yale Law Journal* 1211, 1236 (1949).

19. Chairman Richardson's summary of the loyalty program was given in testimony before a Senate subcommittee on April 5, 1950. See *New York Times*, April 6, 1950, p. 1, col. 5.

20. H. S. Commager, note 13 above, p. 195.

21. The difficulties of obtaining Soviet scientific periodicals is well described in R. Peiss, "Problems in the Acquisition of Foreign Scientific Publications," 22 *Department of State Bulletin* 151 (1950).

22. With the above analysis of the decline of German scientific achievement under the Nazis, compare Vannevar Bush, *Modern Arms and Free Men* (paperbound ed.; Simon & Schuster; copyright, 1949, by The Trustees of Vannevar Bush Trust), pp. 87–88:

"Why were they [the Germans] so far behind [in atomic bomb research]? Bombing and the destruction of needed industrial facilities account for some of the lag. Limited availability of critical materials accounts for some. But the real reason is that they were regimented in a totalitarian regime. There was nothing much wrong with their physicists; they still had some able men in this field in spite of their insane rape of their own universities. They were not as able as they thought they were, or as they probably still think, for their particular variety of conceit is incurable. But they were able enough to have made far greater progress than they did. Their industry certainly demonstrated that it could produce under stress such complicated achievements of science and engineering as the jet plane. Their Fuhrer and their military were certainly keen for new weapons, especially a terror weapon with which to smite England. Yet they hardly got off the starting line on the atomic bomb.

"A perusal of the account of German war organization shows the reason. That organization was an abortion and a caricature. Parallel agencies were given overlapping power, stole one another's materials and men, and jockeyed for position by all the arts of palace intrigue. Nincompoops with chests full of medals, adept at those arts, presided over organizations concerning whose affairs they were morons. Communications between scientists and the military were highly formal, at arm's length, at the highest echelons only, and scientists were banned from all real military knowledge and participation. Undoubtedly the young physicist who penetrated to the august presence of the Herr Doktor Geheimrat Professor said *ja* emphatically and bowed himself out, if he did not actually suck air through his teeth. The whole affair was shot through with suspicion, intrigue, arbitrary power, formalism, as will be all systems that depend for their form and functioning upon the nod of a dictator. It did not get to first base in the attempt to make an atomic bomb."

23. The fullest account of the genetics controversy in the USSR appears in a recent volume by Professor Conway Zirkle, *Death of a Science in Russia* (University of Pennsylvania Press, 1949), in which a large number of documents are collected in a valuable translation. And see also the May 1949 issue of the *Bulletin of the Atomic Scientists* (Vol. 5, No. 5) containing articles by Dunn, Dobzhansky, and others on the suppression of free investigation in genetics in the USSR; also, H. H. Plough, "Bourgeois Genetics and Party-Line Darwinism," 18 *American Scholar* 291 (1949); P. E. Mosely, "Freedom of Artistic Expression and Scientific Inquiry in Russia," 200 *The Annals* 254, 269 *et seq.* (1938). There is, however, some expression of opinion that there is considerable hyperbole in accounts of the "liquidation" of biologists who persist in "bourgeois heresies." For this view, see articles by Marcel Prenant (of the Sorbonne) and Jeanne Lévy (of the Faculty of Medicine in Paris) in *La Pensée*, No. 21, pp. 29, 33 (1948). Translations appear under the titles of, respectively, "The Genetics Controversy" and "Lysenko and the Issues in Genetics," in 13 *Science & Society* 50, 55 (Winter 1948–1949).

24. *Cultural Relations between the United States and the Soviet Union,* State Dept. Publication 3480 (April 1949).

25. Dr. Parin's speech in New York was reported in 4 *American Review of Soviet Medicine* 292, 297 (1947).

Former Ambassador Walter Bedell Smith in his memoirs, *My Three Years in Moscow* (J. B. Lippincott Co., 1950), p. 293, says explicitly, "Dr. Parin, on his return, was tried and sentenced for prematurely revealing the secrets of Soviet Science."

26. The Mundt and Reece statements about the Loyalty Order appear in a round-up of opinion reported in the *New York Times*, March 23, 1947, p. 48, col. 4. The political background of the Loyalty Order is extensively developed in T. I. Emerson and D. M. Helfeld, note 1 above, pp. 8–20.

27. *Report of the President's Temporary Commission on Employee Loyalty* (1947), p. 21.

28. Attorney General Clark's comments on the first 2,000,000 loyalty investigations are reported in the *New York Times*, September 19, 1948, p. 28, col. 4.

29. A brief description of British experience may be found in D. C. Williams, "How Britain Tests Loyalty," *Nation*, November 5, 1949, p. 444. As of the autumn of 1949 some sixty civil servants (out of 100,000) were faced with charges after investigation. About twenty were cleared of the charges. Most of the

others, who either declined to contest the charges or who were not successful in their defense, have simply been transferred to other positions (of equivalent grade) in "non-sensitive" areas.

Chapter VII

1. The authority under which the ONR and the AEC expend research funds is derived from the following statutes:

Section 3 of the Atomic Energy Act of 1946, 60 Stat. 755, 758, directs the Atomic Energy Commission ". . . to insure the continued conduct of research and development activities . . . relating to—(1) nuclear processes; (2) utilization of fissionable and radioactive materials for medical, biological, health, or military purposes" and for "processes entailed in the production of such materials for all other purposes, including industrial uses . . ." The Act itself creates a Division of Research in the AEC, which has set up advisory bodies to help it select the projects for which funds should be made available.

The ONR draws its power from Public Law 588, 60 Stat. 799, 5 U.S.C. § 475 (1946). This statute recognizes that long-range research must be set up on a more solid basis than is possible when annual appropriations must be sought for specific projects of foreseeable utility. The ONR is established to perform duties "relating to the encouragement, promotion, planning, initiation, and coordination of naval research and the conduct of naval research in augmentation of and in conjunction with the research and development conducted by the respective bureaus and other agencies and offices of the Navy Department."

To a much lesser extent than either the ONR or the AEC, the Public Health Service encourages research by nongovernmental institutions and individuals. Its authority derives from the Public Health Service Act, § 301, 58 Stat. 691, 42 U.S.C. § 241, which speaks of ". . . studies relating to the causes, diagnosis, treatment, control, and prevention of physical and mental diseases and impairments of man. . . ."

2. The AEC's program of university research, especially in the biological sciences, is discussed in the *Sixth Semiannual Report* (July, 1949), pp. 112 *et seq.* and 161 *et seq.* And see J. E. Pfeiffer, "The Office of Naval Research," 180 *Scientific American*, No. 2, p. 11 (1949); A. T. Waterman and R. D. Conrad, "The Office of Naval Research," 16 *American Scholar* 354 (1947).

3. An example of studies only remotely military supported by the AEC and ONR is found in a University of Rochester project reported by L. E. Young *et al.* in "Hemolytic Disease in Newborn Dogs Following Isoimmunization of the Dam by Transfusions," 109 *Science* 630 (1949). Current jointly sponsored projects include cloud-chamber cosmic ray studies at the University of Washington, Beta-ray spectra studies at the University of Southern California, and research in radiobiology and chemical genetics at Amherst College. As of April 1950 the AEC was supporting 67 nonsecret research projects in the physical

sciences, and in addition shared with the ONR in supporting another 60 non-secret physical research projects. The 127 projects were scattered among 83 different institutions. AEC Release No. 280, April 26, 1950.

4. Dr. Pitzer's remarks on the subject of keeping atomic energy research out of university laboratories occurred in the course of a press conference, reported in the *New York Times,* January 19, 1949, p. 29, col. 4.

5. The remarks about the transference of scientific ideas are quoted from J. D. Bernal, *The Social Function of Science* (George Routledge & Sons, Ltd., 1939), p. 303.

6. *The Hoover Commission's Task Force Report on Public Welfare,* Appendix P, pp. 557–590, accumulates federal expenditures for educational purposes, including research. The notations for research for fiscal 1949 total $204,713,000.

7. Dr. Day's remarks are quoted from a memorandum from him to Dean C. C. Murdock, Dean S. C. Hollister, and Vice President T. P. Wright, entitled "Policy Relating to Classified Research on the Campus," September 9, 1948.

8. The experience of the German academic community has been touched upon by Leo Szilard, "The AEC Fellowships: Shall We Yield or Fight?", 5 *Bulletin of the Atomic Scientists* 177 (1949):

"A few months after the Hitler government was installed in office, it demanded that instructors of the Jewish faith be removed from their university positions. At the same time, every assurance was given that professors who had tenure would remain secure in their jobs.

"The German learned societies did not raise their voices in protest against these early dismissals. They reasoned that there were not many Jewish instructors in German universities anyway, and so the issue was not one of importance. Those of the dismissed instructors who were any good, so they pointed out, were not much worse off, since they were offered jobs in England or America. The demand of the German government for the removal of these instructors did not seem altogether unreasonable, since they couldn't very well be expected wholeheartedly to favor the nationalist revival which was then sweeping over Germany. To the learned societies it seemed much more important at that moment to fight for the established rights of those who had tenure, and this could be done much more successfully, so they thought, if they made concessions on minor points.

"In a sense the German government kept its word with respect to those who had tenure. It is true that before long most professors who were considered 'undesirable' were retired; but they were given pensions adequate for their maintenance. And these pensions were faithfully paid to them until the very day they were put into concentration camps, beyond which time it did not seem practicable to pay them pensions. Later many of these professors were put to death, but this was no longer, strictly speaking, an academic matter with which the learned societies needed to concern themselves."

9. The play of politics upon education in Germany is described in E. Y.

Hartshorne, "The German Universities and the Government," 200 *The Annals* 210, 223 (November 1938): In the four years after the National Socialist Party came to power, 44 per cent of all the teachers in the natural science faculties of German universities disappeared from the faculty rosters; in the medical faculties, 48.8 percent of the teachers were dropped. The percentage increase over the loss of professors in the preceding four years in these two groups was, respectively, 185 and 218 per cent. Hartshorne says that "there can be no doubt that the German scientists who were ousted from their University positions by National Socialist legislation were attacked not as chemists or historians or mathematicians, and so forth, but as 'State enemies' according to the official definition, in terms of ancestry or political viewpoint."

10. President J. B. Conant's remarks about the qualifications for joining a community of scholars are quoted from *The President's Report, 1947, Harvard University*, pp. 3, 4.

11. The military profession's traditional coolness toward the novelties of science has been touched upon by Waldemar Kaempffert in "Science, Technology, and War" in *Civil-Military Relationships in American Life* (ed. Kerwin; University of Chicago Press, 1948), pp. 14, 16: "An army is a highly organized and planned artificial society. . . . Since war is a matter of life and death, victory or defeat, it might be supposed that new death-dealing inventions would be eagerly sought. But innate opposition to change is as inherent in soldiers as it is in financiers and industrialists. . . . There was less cultural lag in science during the war recently ended than in any of its predecessors. The reason is to be found in the way research was organized. In the past the military dominated research. Even before war was declared on the Axis powers, President Roosevelt had created the National Defense Research Committee, later merged into the Office of Scientific Research and Development. Civilian scientists sat with representatives of the Army and Navy on various boards, but the civilians outnumbered the military. Hence the boldness of thinking and experimentation . . ."

For relevant comments, see Vannevar Bush, *Modern Arms and Free Men* (Simon & Schuster, 1949), pp. 26, 33, 61.

12. The quotation of General Sir Ian Hamilton is from his book, *The Soul and Body of an Army* (Edward Arnold & Co., 1921), pp. 61–62.

13. Enrico Fermi's comment on free choice of research projects is quoted by M. Polanyi in "The Foundations of Freedom in Science," 2 *Bulletin of the Atomic Scientists* 6 (December 1946).

14. The National Science Foundation was created by Public Law 507, 81st Cong., 2d Sess., which became law on May 10, 1950. Among its other duties, the Foundation is authorized and directed "to initiate and support basic scientific research in the mathematical, physical, medical, biological, engineering, and other sciences, by making contracts or other arrangements (including grants, loans, and other forms of assistance) . . ." When this book went to press, the Foundation was not yet organized; its National Science Board and Director

had not been appointed and appropriations of funds had not been made. In due time, however, it was anticipated that the Foundation would be the sponsor of most of the nonsecret research now supported by the Federal Government through the ONR and the AEC. The case for the National Science Foundation is stated, among other places, in Vol. I of *Science and Public Policy*, A Report to the President by John R. Steelman, Chairman, the President's Scientific Research Board, pp. 31–35 (1947). And see also the extended record of hearings, covering nearly a thousand pages, on proposed science legislation, embodied in S. 1297 and related bills, before the Senate Committee on Military Affairs, 79th Cong., 1st Sess. (1945).

15. Dr. Gregg's remarks about researchers are quoted from his article, "We Must Not Lag in Medical Research," *New York Times Magazine*, August 7, 1949, p. 13, at 74.

16. The manpower problem in American science is extensively considered in the Steelman report, cited in note 14 above, at pp. 14–23.

17. The most extensive descriptive material in print concerning the Atomic Energy Commission Fellowship Program is probably to be found in the record of hearings on that subject before the Joint Committee on Atomic Energy, 81st Cong., 1st Sess., May 16, 17, 18, and 23, 1949. A list of the AEC fellowships for 1949–1950 appears in the AEC's *Sixth Semiannual Report* (July 1949), pp. 183–189. A description of the fields of specialization of the fellows appears in the record of the hearings, above, at pp. 181–197.

18. The relationship of AEC fellowship projects to the AEC's other concerns is indicated in the testimony of Detlev W. Bronk, chairman of the National Research Council, before the Joint Committee on Atomic Energy, *Hearing on Atomic Energy Commission Fellowship Program*, 81st Cong., 1st Sess. (1949), pp. 80–81.

19. The composition of the fellowship boards set up by the NRC to pass on AEC fellowship applications is shown in the AEC's *Sixth Semiannual Report* (July 1949), pp. 183 ff.

20. At the end of 1949, research work was being carried on in Brookhaven National Laboratory by fifteen graduate students; at the Argonne National Laboratory, by ten; and at Oak Ridge, by seven. These were not necessarily fellowship recipients.

21. The quoted reasons for the AEC's policy as to clearing fellows are set forth in a letter from the Commission to Senator Hickenlooper, dated October 11, 1948, printed in the record of the hearing cited in note 18 above at p. 7.

22. Dr. Richards' and Dr. Bronk's remarks about educating a Communist appear in the record of the same hearing at pp. 14, 72–74.

23. Dr. Oppenheimer's comment upon the sources of great discoveries appears in the same record at p. 89.

24. Dr. Conant's objections to widespread investigations are recorded in the same volume at p. 159.

25. Dr. DuBridge's sentiments about investigating students were expressed

in a hearing before the Joint Committee on Atomic Energy, *Investigation into the United States Atomic Energy Project,* 81st Cong., 1st Sess. (July 8, 1949), pp. 848, 849.

26. The views of the American Institute of Physics concerning investigations of AEC fellows are reflected in a telegram from five of its board members, addressed to Senator McMahon, reported in the *New York Times,* May 27, 1949, p. 10, col. 2. The signatories were George R. Harrison, editor of the *Journal of the Optical Society of America;* Paul E. Klopsteg, director of the Northwestern University Technological Institute; F. W. Loomis, head of the Physics Department of the University of Illinois; George B. Pegram, vice president of Columbia University; and Wallace Waterfall, secretary to the governing board of the Institute.

27. The law regarding AEC fellowship funds is found in Sec. 102-A of the Independent Offices Appropriation Act, Public Law No. 266, 81st Cong., 1st Sess. (1949). The section was added to the Act as a rider during the Senate's consideration of the appropriation bill. It was debated in the Senate on August 2, 1949 (95 *Cong. Rec.* 10822–10829); but when it came up in the House, even a reading of the text of the rider was dispensed with and the measure was adopted by unanimous consent, August 15, 1949 (95 *Cong. Rec.* 11739).

28. Dr. Richards' description of fellowship applicants is quoted from his testimony before the Joint Committee on Atomic Energy, *Hearing on Atomic Energy Commission Fellowship Program,* 81st Cong., 1st Sess. (1949), p. 14.

29. The results of the investigations of 151 fellows are touched upon in a colloquy between Senator Knowland and Dr. Shields Warren, director of the Commission's Division of Biology and Medicine, in the record of the same hearing, at pp. 16–17.

30. In connection with estimates of the dimensions of the "loyalty problem" among fellowship applicants, it may be worth recording that the AEC, without awaiting an explicit command from Congress, receded from its original position because of the furor created by the North Carolina Communist's fellowship. It decided early in the summer of 1949 to require all fellows to execute a loyalty oath and affidavit; moreover, it decided to require in the future that a check be made of existing FBI records. During the summer of 1949 the first group of fellows, numbering 497, were called on to execute the prescribed loyalty oaths and non-Communist affidavits. The North Carolina Communist refused as did two others. Whether the two others refused because they could not subscribe the oaths or because of opposition to them in principle is not known. The three fellowships, out of these 497, were thereafter withdrawn.

31. Senator Hickenlooper's belief about what the American people will stand appears in Joint Committee on Atomic Energy, *Hearing on Atomic Energy Commission Fellowship Program,* 81st Cong., 1st Sess. (1949), p. 65. At p. 66 Senator Hickenlooper differentiated between the fellowship program and other federal-aid-to-education programs. The latter, he observed, involved aid to educational institutions, rather than particular individuals. While it is true

that a Communist might benefit from a school-aid measure, indeed might even be compelled to benefit by virtue of compulsory attendance laws, still "it is not a specific subsidy to the individual."

32. Congressman Durham's related remarks are quoted from the same record, at p. 104.

33. Senator Millikin's comments were made at a subsequent hearing before the Joint Committee on Atomic Energy, *Investigation into the United States Atomic Energy Project*, 81st Cong., 1st Sess. (1949), pp. 850–851.

Lieutenant General Walter Bedell Smith reports that a similar drift of thought has been brought to its logical culmination in the Soviet Union. He tells us that in the USSR "Higher Education is reserved for those who develop a 'political consciousness' to a very high degree or for the offspring of the new Soviet aristocracy." *My Three Years in Moscow* (J. B. Lippincott Co., 1950), p. 114.

34. The exchange between Senator Hickenlooper and Dr. DuBridge concerning AEC fellowships appears in the last-cited hearings at p. 853. And see also p. 850. Senator Hickenlooper had made a similar analysis of the G.I. bill of rights on an earlier occasion, saying that the educational opportunities provided under that law "are considered to be an earned matter of right, which has already been earned and paid for. The consideration has been given for that. Under the fellowship program it is a matter of governmental grace. It is extending a gratuity that is not considered to be an earned award, except as such gratuities may develop some potentials." *Hearing on Atomic Energy Commission Fellowship Program*, p. 60.

35. Dr. Oppenheimer's explanation of the reason for a fellowship program appears in the last-cited hearing record, at p. 89. Dr. Conant's related opinion appears at p. 159, and Dr. Gregg's at p. 93.

36. Some of the expressions of distinguished academic persons on the subject of submitting to oaths and investigations are perhaps worth recording.

Enrico Fermi, *Investigation into the United States Atomic Energy Project*, Hearing before Joint Committee on Atomic Energy, 81st Cong., 1st Sess. (1949), p. 868: ". . . to be considered a poor risk is no irrelevant matter for a young man who has not had a chance to be established. A young man who is trying to acquire that competence that may eventually lead him into establishing himself may properly object to the danger of being so branded. I believe that the percentage of those who would be properly weeded out by a loyalty investigation is extremely small, but I believe that a widespread investigation of students not engaged in secret work would have a very depressing influence."

Lee A. DuBridge, *ibid.*, pp. 855, 859: "The harm comes from the very considerable number of perfectly honest and loyal men who will be disqualified on evidence which is quite inconclusive and possibly even wrong.

"Now, during the war I saw many people, honest and loyal men, disqualified for employment in war programs on misunderstandings, on incomplete information, on misunderstood information, on all sorts and kinds of information, some of which might be true and some of which the truth could not be verified. There will be a large number of perfectly loyal and able young Americans who will be disqualified. That will be a very serious matter to them.

"How many of these people who are said to be Communists or fellow travelers—and, incidentally, the terms are not very clearly defined—simply have a sort of political feeling usually on the basis of youthful naivete that here is something new they wish to explore because it has some attractive features?

"They will get into it and get sick of it in a couple of years and be out of it and be better and more loyal Americans perhaps as a result of their practical introduction to communism and the Communist conspiracy. It is not as though these people are permanently tied up and are forever subversive citizens. They may not be. If they are, they should be investigated by the FBI, and they should be brought through the normal legal procedures for necessary punishment for any illegal or subversive action. But let's not extend police-state methods to a large section of American life in the hysterical fear that one or two such fellows are going to overthrow the country. I do not think they will."

James B. Conant, Joint Committee on Atomic Energy, *Hearing on Atomic Energy Commission Fellowship Program*, 81st Cong., 1st Sess. (1949), p. 159: ". . . there is always a good chance that, as in the past, a certain number of young Communists would have a revulsion of feeling and leave the party. I trust Congress will not take any action which will of necessity involve proceedings creating an atmosphere of distrust and suspicion in the scientific world as I feel certain the loss to the country will far outweigh the possible hazards involved in the calculated risk of the method now used."

J. R. Oppenheimer, *ibid.*, pp. 89–90: "They [i.e., 'loyalty' procedures] involve secret, investigative programs which make difficult the evaluation and criticism of evidence; they take into consideration questions of opinion, sympathy, and association in a way which is profoundly repugnant to the American tradition of freedom; they determine at best whether at a given time an individual does have sympathy with the Communist program and association with Communists, and throw little light on the more relevant question of whether the man will in later life be a loyal American. It would be foolish to suppose that a man against whom no derogatory information can be found at the age of 20 was by virtue of this guaranteed loyal at the age of 30. It would be foolish to suppose that a young man sympathetic to and associated with Communists in his student days would by that fact alone become disloyal, and a potential traitor. It is basic to science and to democracy alike that men can learn by error.

"My colleagues [on the General Advisory Committee to the AEC] and I attach a special importance to restricting to the utmost the domain in which special secret investigations must be conducted. For they inevitably bring with them a morbid preoccupation with conformity, and a widespread fear of ruin, that is a more pervasive threat precisely because it arises from secret sources. Thus, even if it were determined, and I do not believe that it should be, that on the whole the granting of fellowships, or, more generally, of Federal support to Communist sympathizers, were unwise, one would have to balance against this argument the high cost in freedom that is entailed by the investigative mechanisms necessary to discover and to characterize such Communist sympathizers. This is what we all have in mind in asking that these intrinsically repugnant security measures be confined to situations where real issues of security do in fact exist and where, because of this, the measures, though repugnant, may at least be intelligible."

A. Newton Richards, *ibid.*, p. 16: "The effect of a disqualification by AEC of a successful applicant on political grounds, would be a stigma—a wound—which would be apt never completely to heal—and it would be given at an age at which it could well produce the greatest degree of damage. It is intolerable to me to think that in order to assure our taxpayers that their money is not being spent in the training of Communists, we must subject a most precious part of our intellectual resources to the risk of hurt involved in such investigation, where no question of security is involved."

37. The words quoted relating to the effect of investigations upon educational atmospheres and the outlook of youth are from a statement, endorsed by all members of the AEC's General Advisory Committee, signed by them June 6 and published August 5, 1949. It appears in 110 *Science* 197 (August 19, 1949). In part it reads as follows: ". . . we are horrified by the prospects of moving this whole semi-police apparatus into the realm of youth. We believe that the reputation of many young people of the country might be . . . impaired by rumors growing out of such a system of investigation of prospective fellowship holders. Older people can see in proper perspective calls from FBI agents, they can answer questions about acquaintances without feeling that the man being investigated is under suspicion. But young people of university age are likely to react quite differently. An atmosphere of suspicion and uncertainty is likely to be generated by the activities of federal agents among many groups of friends in colleges, universities, and in local communities. In short, the results of requiring investigations of candidates for fellowships will have serious repercussions throughout the country. . . ."

And see also the letter of Marshall Stone, chairman of the Department of Mathematics at the University of Chicago, resigning from the NRC's Postdoctoral AEC Fellowship Board for Mathematics, Physics, and Chemistry. The letter appears in 110 *Science* 191 (August 19, 1949).

38. Senator McMahon's remarks about the intellectual qualifications of the North Carolina Communist were made before the Joint Committee on Atomic Energy, *Hearing on Atomic Energy Commission Fellowship Program*, 81st Cong., 1st Sess. (1949), pp. 60–61.

39. Senator Knowland's description of "the calculated risk" will be found in the record of the same hearing at p. 43.

40. Senator Millikin's statements regarding the expense involved in the investigation of AEC fellows were made at the hearing before the Joint Committee

on Atomic Energy, *Investigation into the United States Atomic Energy Project,* 81st Cong., 1st Sess. (1949), p. 852.

41. The views of the National Academy of Sciences, and of the National Research Council which is its adjunct, were embodied in a statement drawn up at Academy meetings on October 24–26, 1949, communicated to the Commission on November 2, 1949, in a letter from A. N. Richards, the Academy's president, to Carroll Wilson, the AEC's General Manager.

42. The 1950–1951 fellowship program was described by the AEC in its Release No. 236, December 16, 1949. Copies of the correspondence between the Commission and the National Academy of Science are annexed to this same release. More recently the AEC has initiated an additional small program, involving up to four fellowships per year, for training in industrial medicine immediately related to work at atomic energy installations. All candidates must be fully investigated and must have a security clearance. AEC Release No. 292, June 25, 1950.

43. The AEC's new predoctoral fellowship program is described in AEC Release No. 256, February 12, 1950. The administering bodies are Associated Universities, Inc.; Oak Ridge Institute of Nuclear Studies, Inc.; a board appointed by the Argonne National Laboratory; and a board established by the University of California.

44. Official correspondence concerning the National Science Foundation amendment and the stated reasons for its ultimate rejection by the Congress can be found in the *Conference Committee Report to Accompany S. 247,* 81st Cong., 2d Sess., House Rep. No. 1958 (1950), pp. 13 ff.

Chapter VIII

1. The policy of withholding information about the sources of FBI information long antedates the Loyalty Order and is not by any means confined to loyalty or personnel security cases. For example, in *United States* v. *Kohler Co.,* 18 *U.S. Law Week* 2035 (E. D. Pa., June 28, 1949), it was held that a defendant in an antitrust action was not entitled to learn what the FBI had been told by customers and others with whom the defendant had had business dealings. But in a case like that, unlike the ones we are now discussing, the stories told to the FBI could not be used as evidence against the defendant unless they were repeated in open court.

The policy of nondisclosure was explained in the following terms in 40 Op. Atty-Gen., No. 8, pp. 46, 47, April 30, 1941:

"Disclosure of the reports would be of serious prejudice to the future usefulness of the Federal Bureau of Investigation. As you probably know, much of this information is given in confidence and can only be obtained upon pledge not to disclose its sources. A disclosure of the sources would embarrass informants—sometimes in their employment, sometimes in their social relations, and in extreme cases might even endanger their lives. We regard the keeping of faith with confidential informants as an indispensable condition of future efficiency."

2. General Donovan's views are embodied in an article by him and Mary Gardiner Jones, "Program for a Democratic Counter Attack to Communist Penetration of Government Service," 58 *Yale Law Journal* 1211, 1235 (1949).

And see also John Lord O'Brian, "Loyalty Tests and Guilt by Association," 61 *Harvard Law Rev.* 592 (1948).

3. The case of the employee who had a chance to face his accusers is discussed in *Senate Rep. No. 1169,* 81st Cong., 1st Sess. (1949), pp. 59, 68.

4. Mr. Clark's comment on gossip in FBI reports appears in the *New York Times,* July 2, 1949, p. 2, col. 7.

As to the extent of this accumulation of gossip, a former Attorney General, Homer S. Cummings, has revealed that as early as 1919 the "General Intelligence Division" of the FBI, under J. Edgar Hoover's direct supervision, had compiled "suspect indexes" containing some 200,000 names of persons who were thought to have engaged in "radical activities." Mr. Hoover subsequently acknowledged that the "counterradical" activities which were conducted under his direction were not at that time within the Department of Justice's jurisdiction "as there has been no violation of the federal laws." Homer S. Cummings and Carl McFarland, *Federal Justice* (Macmillan, 1937), pp. 429, 430. One can only guess to what astronomical heights the number of dossiers has risen since 1919.

5. The Supreme Court's words about the unreliability of unprobed testimony are quoted from *Eccles* v. *Peoples Bank,* 333 U.S. 426, 434 (1948). In a case in which the plaintiff's claims of injury were supported entirely by affidavits, the Court refused to issue a declaratory judgment, saying: "Judgment on issues of public moment based on such evidence, not subject to probing by judge and opposing counsel, is apt to be treacherous   Modern equity procedure has tended away from a procedure based on affidavits and interrogatories, because of its proven insufficiencies. . . ."

6. Mr. Jackson on April 30, 1941, declined to furnish certain FBI reports to the House Committee on Naval Affairs, saying: "Disclosure of information contained in the reports might also be the grossest kind of injustice to innocent individuals. Investigative reports include leads and suspicions, and sometimes even the statements of malicious and misinformed people. Even though later and more complete reports exonerate the individuals, the use of particular or selected reports might constitute the grossest injustice, and we all know that a correction never catches up with an accusation." 40 Op. Atty-Gen., p. 47 (1941).

7. On the judicial attitude toward informers' testimony see, e.g., *District of Columbia* v. *Clawans,* 300 U.S. 617, 630 (1937); *Colyer* v. *Skeffington,* 265 Fed. 17, 69 (D. Mass., 1920, opinion by Anderson, C.J.), reversed on other grounds

in 277 Fed. 129 (C.C.A. 1st, 1922); *Todd* v. *State*, 246 Pac. 492 (Okla. Crim. App., 1926); *People* v. *Loris*, 131 App. Div. 127, 130 (2d Dept., 1909); *State* v. *Bryant*, 97 Minn. 8, 10 (1905).

8. Occasionally the informing spirit has been officially inspired, as in recent years in Hungary, Russia, and elsewhere. Even in our own country the stimulation of the informing spirit is not unknown, as witness the request by then Police Commissioner Toy to the people of Detroit on July 7, 1949: "If anyone suspects a city employee of being a Communist or a Red sympathizer, we would welcome the information, either signed or anonymous. Those signing the letters will be protected." Historical consequences of unbridled denunciations may be studied in E. S. Mason, *The Paris Commune* (Macmillan, 1930), p. 286, and in Walter Duranty, *USSR* (J. B. Lippincott Co., 1944), p. 223. Duranty, describing the sequel of the Soviet "treason trials," says, "The clouds of doubt and anxiety became a storm of frenzy and hysteria, until no man knew whom to trust, and children denounced their parents, brother attacked brother, and husband accused wife. The 'Great Purge,' as it was called, raged for nearly two years, from 1936 to 1938, and caused vast confusion, disorganization and distress at the very time when Stalin was doing his utmost to prepare his country for war."

9. There has not yet been a great deal of litigation in which loyalty proceedings have been challenged. In one case, *Washington* v. *Clark*, 84 F. Supp. 964, 967 (D. District of Columbia, June 28, 1949), Judge Holtzoff wrote: "If the requirements of due process laid down by the Fifth Amendment of the Constitution of the United States were applicable to the discharge of a Government employee from the service, this order would not comply with those requirements. . . . But the requirements of the due process clause . . . do not apply to the employer-employee relationship as between the Government and its employees. An employer does not have to grant to his employee a formal trial, with all its pomp and circumstance. . . ."

This curious judicial characterization of due process as being "pomp and circumstance" appears to have influenced the same judge's thinking in *Bailey* v. *Richardson*, Pike-Fischer *Administrative Law* 31c. 223–7 (D. District of Columbia, July 27, 1949), in which he wrote: "The only difficulty from the plaintiff's standpoint is that she was not confronted with the evidence against her, and was not informed of the names or the identity of the witnesses who gave the information against her. We must bear in mind, however, that a Government employee is not entitled to an open trial under the Constitution . . . The Court realizes that it is most unusual not to disclose to a person the evidence against him, but . . . there is no legal or constitutional right in plaintiff to have a hearing of any kind."

The District of Columbia Court of Appeals affirmed the judgments adverse to Washington and to Bailey on April 17 and March 22, 1950, respectively. The court's opinions are not yet officially reported, but the *Bailey* case appears in

Pike-Fischer *Administrative Law* 31c.224. The Supreme Court granted certiorari in *Bailey* v. *Richardson* on June 5, 1950, 70 Sup. Ct. 1025.

For other cases, not involving loyalty issues, manifesting a similar judicial unwillingness to intervene in matters affecting a federal employee's status, see *Carter* v. *Forrestal*, 175 F. 2d 364 (App. D.C., 1949), and cases there cited; Howard C. Westwood, "The 'Right' of an Employee of the United States against Arbitrary Discharge," 7 *George Washington Law Review* 212 (1938).

10. Felix Frankfurter's comments about stating the grounds of one's conclusions appear in his summary of the Cincinnati Bar Association Conference on Functions and Procedure of Administrative Tribunals, in 12 *University of Cincinnati Law Review* 117, 260, 276 (1938).

11. The views of the Attorney General's Committee on Administrative Procedure on the subject of findings and decisions are embodied in its *Final Report*, 78th Cong., 1st Sess., Senate Doc. No. 8 (1941), p. 30. In the same section of its report the Committee emphasized that "the exposure of reasoning to public scrutiny and criticism is healthy. An agency will benefit from having its decisions run a professional and academic gauntlet." Moreover, "the parties to a proceeding will be better satisfied if they are enabled to know the bases of the decision affecting them. Often they may assign the most improbable reasons if told none. Finally, opinions enable the private interests concerned, and the bar that advises them, to obtain additional guidance for their future conduct."

And see also J. P. Chamberlain, N. T. Dowling, P. R. Hays, *The Judicial Function in Federal Administrative Agencies* (Commonwealth Fund, 1942), pp. 60–61.

12. The Supreme Court's emphasis on the need for clear findings is reflected in *Colorado-Wyoming Gas Co.* v. *Federal Power Commission*, 324 U.S. 626, 634 (1945); and see also *Beaumont, S. L. & W. Ry. Co.* v. *United States*, 282 U.S. 74 (1930); *Florida* v. *United States*, 282 U.S. 194 (1931); *Yonkers* v. *United States*, 320 U.S. 685 (1944).

13. The courts' attitude toward the especial desirability of findings when there are several possible bases of administrative decision is well illustrated by *Niagara Frontier Co-operative Milk Producers Bargaining Agency* v. *Du Mond*, 297 N.Y. 75, 74 N.E. 2d 315 (1947): *United States* v. *Chicago, M., St. P. & P. R. Co.*, 294 U.S. 499 (1935); *Elite Dairy Products, Inc.* v. *Ten Eyck*, 271 N.Y. 488, 3 N.E. 2d 606 (1936); *Jacob Siegel Co.* v. *Federal Trade Commission*, 327 U.S. 608 (1946).

14. The suggestion that loyalty and security decisions should be accompanied by an indication of reasoning finds an interesting parallel in the "Canons of Judicial Ethics," published in 54 *Reports of American Bar Association* 927 (1929). Canon 19, dealing with judicial opinions, states: "In disposing of controverted cases, a judge should indicate the reasons for his action in an opinion showing that he has not disregarded or overlooked serious arguments of coun-

sel. He thus shows his full understanding of the case, avoids the suspicion of arbitrary conclusion, promotes confidence in his intellectual integrity and may contribute useful precedent to the growth of the law."

Even in connection with patent applications and many similar administrative matters of considerably less moment than a human career, there are legal requirements that opinions be supported by reasons. See, e.g., *Notice Required on Rejection of a Patent Claim*—35 U.S.C.A. § 51 (1940); *International Standard Electric Corp.* v. *Kingsland*, 169 F. (2d) 890, 892 (App. D.C., 1948).

15. The AEC is fully aware that newcomers do not have the procedural protections accorded "old hands." In its *Fourth Semiannual Report* (July 1948), p. 15, the AEC announced: "Whether hearings should be granted to applicants for employment, as they are for persons already employed on atomic energy work, is a matter currently under consideration. . . ." In its *Fifth Semiannual Report* (January 1949), p. 123, the AEC said: "As the Fourth Semiannual Report states, the Commission is studying the desirability of granting hearings to applicants for employment who have been denied clearance." Frequent inquiries concerning the progress of these studies have been given the laconic response, "Still studying." As a matter of fact, in a few exceptional cases hearings have been accorded applicants. No pattern or policy seems to have been reflected in these cases; for all that appears, they were merely instances in which the applicant's friends or attorneys were sufficiently exalted to obtain special consideration.

16. The figures on AEC formal denials were supplied the author by the Commission. The estimate of the number of scientists who had not received clearance was given to the press by Dr. William A. Higinbotham, associate director of the Electronics Division of Brookhaven. *New York Times,* May 29, 1949, p. 1, col. 3.

17. The Committee report that discusses clearance statistics is the *Report on Investigation into the United States Atomic Energy Commission*, 81st Cong., 1st Sess., Senate Rep. No. 1169 (1949), p. 66.

18. The comments upon the difference between a file review and a hearing review are derived from an address delivered on September 21, 1949, by Dr. John A. Swartout, as part of a Symposium on Security Clearance and the Scientist, during the 116th national meeting of the American Chemical Society. The full paragraph from which phrases have been quoted reads as follows: "In the course of reading the security files on such individuals, one is overwhelmed by the succession of testimony by witnesses and the accumulation of information which combine to set a pattern pointing to the unreliability of the suspect. How misleading such an accumulation of information can actually be is best illustrated by another reference to the 'average case.' During consideration of questionable cases, the files are usually reviewed by a number of the AEC security force. Their comments and conclusions are recorded. In a few abnormal cases, more than one 'hearing' has been held either before the local board or in a more informal session with security officials. It is very pertinent

that, when a review is based only on the file, the decision is, as a rule, against the granting of clearance, whereas the hearings in which the sources of information are confronted nearly always have reversed this decision completely." Dr. Swartout's address has been printed in 5 *Bulletin of the Atomic Scientists* 337 (1949).

19. The question of wartime experience with cases of mistaken identity is touched on in Irvin Stewart, *Organizing Scientific Research for War* (Little, Brown & Co., 1948), p. 30.

20. The figures about discontinuance of FBI investigations because of mistaken identity are given in W. J. Donovan and M. G. Jones, note 2 above, pp. 1229–1230.

Officials of Carbide & Carbon Chemicals Corporation, the operator of Oak Ridge, stated in an interview with the author that they know of FBI reports that are shot through with erroneous allegations and mistakes in identity. For this reason they would support, though they would not themselves urge initiating, a policy of granting hearings to applicants. One experienced officer said that there was frequent confusion between a man and his similarly named relatives. Another said that especially in cases involving Negroes there is a tendency on the part of sheriffs and others who give information to the FBI to make erroneous identifications.

21. J. Edgar Hoover's remarks about mistaken identity are to be found in a guest column he wrote for the *New York Daily Mirror*, June 22, 1949, p. 4, col. 1, which reads in its pertinent parts as follows:

"The primary responsibility of good law enforcement is to protect the welfare of the community. This entails not only detection of the evil-doer, but exoneration of the falsely accused.

". . . Each complaint or bit of information received is thoroughly checked to its ultimate source, and many times our inquiries reflect that the data received is incorrect or the wrong person is involved.

"Recently, in connection with the loyalty of government employes program, we received allegations that a government employe, who also was a member of a labor union, might be a Communist Party member. Investigation revealed that an individual with the same first and last names, but a different middle name, was a Communist, but the government employe was a different individual.

"In another case, the FBI got derogatory information about an individual, whose father and mother, living in a West Coast city, were allegedly members of the Communist Party. Investigation showed, however, that the individual was the son of parents, with identical first and last names, from an Eastern city.

"The FBI was able, in both of these cases, through careful investigation, to ascertain the facts—and thereby to keep unsullied the reputations of two Americans."

22. The Army's difficulties with the question of "employability" may be traced through a succession of newspaper dispatches. The initial disclosure of

the ruling in the Clapp case was in a report by Jack Raymond from Frankfort, appearing in the *New York Times,* June 10, 1949, p. 1, col. 3. Later stories appear in the *Times* of June 11, p. 1, col. 4; June 15, 1949, p. 1, col. 2; June 16, 1949, p. 4, col. 3. In a letter to the American Civil Liberties Union dated July 15, 1949, Secretary of the Army Gordon Gray insisted that the characterization of Roger Baldwin and others as "unemployable" had not been done by a responsible Army source, and said that "instructions have been given to remove from the records any reference to these or other persons as unemployable by the Army and that the fact that such term may have been used casts no reflection upon the persons concerned. . . ." He added that there would be no prejudice against any of these people because of the listing in the event their services were needed in the future.

23. In connection with the AEC's loss of personnel through failure to accord a hearing to applicants, the Joint Congressional Committee on Atomic Energy has this to say in *Senate Rep. No. 1169,* 81st Cong., 1st Sess. (1949), p. 64: "Another [file] relates to a person who never became a Commission employee. He applied for a job and was rejected as a security risk on the basis of his associations, whereupon he renounced any desire to serve with the Commission but demanded a hearing and full loyalty appraisal as a means of exonerating his name. Normally the Commission reserves the benefits of its security review procedure to actual employees about whom a question has arisen, excluding job applicants. In this case, however, the individual believed that his friends and associates knew why he had been rejected; that his chances of securing employment elsewhere were bound to suffer; and that his damaged reputation entitled him to a clean-cut, official finding. Under the circumstances the Commission made a special exception and appointed a local board. After a hearing and evaluation by ranking AEC security officers, the individual was finally determined to be eligible for clearance assuming that he were an employee; and thus he succeeded in removing the original imputation of disloyalty.

"AEC witnesses informed the committee that applicants present a puzzling problem: If the Commission or a contractor desires to hire them, they must be encouraged to mark time for 2, 3, 4, 5, or even 6 months without accepting other regular employment, while the FBI completes an investigation; then, if the investigation means that they cannot be hired for security reasons, the encouragement previously given causes them to make inquiries and often, with the help of rumors and gossip, to glean the truth; in that event they are apt to press tirelessly for a full explanation and an opportunity to clear themselves. The committee itself knows of at least one eminent scientist who refuses to seek Commission employment for fear that, if rejected on security grounds, he could not—as a mere applicant—be permitted a local board hearing and a chance to confront accusers who may be listed in his FBI file."

Chapter IX

1. The Supreme Court's observations about coercive elimination of dissent and the freedom to differ are in *West Virginia State Board of Education* v. *Barnette,* 319 U.S. 624, 640–641, 642 (1943).

2. The Bostonian whose words are quoted was Thomas Brattle, a leading citizen. While the Salem witchcraft frenzy was still alive, he wrote a circular letter analyzing the trials and the evidence adduced. His letter is quoted by Marion L. Starkey, *The Devil in Massachusetts* (Knopf, 1949), pp. 224–225.

Acknowledgments

IN a book of this sort, in which interpretation and opinion play so large a part, no one but the author himself can be assigned responsibility for the product.

I would be less than grateful, however, if I failed to acknowledge the very great help I have received from numerous sources. A grant from the Rockefeller Foundation made it possible to travel widely for the purpose of personally interviewing scientists and administrators in many quarters. Altogether I have had the benefit of the information and ideas of 141 different individuals, scattered from the District of Columbia to the Pacific Coast and from Tennessee to Massachusetts. They have included Government officials, military personnel, staff members of large and small laboratories, industrial scientists, university professors, and practicing attorneys among others. I do not identify them by name because it is not feasible to ascribe specifically to each one the data I obtained from him, and I desire to avoid any embarrassment that might conceivably flow from incorrect assumptions concerning what I learned from whom.

During an early stage of my study I had the pleasure of serving as consultant to a special committee of the American Association for the Advancement of Science that dealt with the civil liberties of scientists. The brief relationship was a stimulating one for me. I have drawn at a few places upon the committee's report, which I helped prepare. The members of that committee were Maurice B. Visscher, chairman, Philip Bard, Robert E. Cushman, R. D. Meier, and James R. Newman.

My colleagues in the project of which the present study is but a unit—the Cornell Research in Civil Liberties—have been generous in reading the manuscript and in making a number of exceedingly helpful suggestions. My immediate associates have been Robert E. Cushman of Cornell, the director of the project; Eleanor Bontecou of Washington, D.C.; and Robert K. Carr of Dartmouth. I cannot overstate my appreciation of their aid, though I entirely exonerate them of blame for the final product.

Jerold M. Lowenstein, Sara Spiro, and Tobias Weiss effectively assisted from time to time in various researches that entered into this undertaking. I am much indebted to all of them. Finally, though by no means least, I am warmly grateful to Teresa B. Sweeney and Mina Sweeney, mother and daughter, who patiently and ably rendered the secretarial services that are probably the most taxing part of a study like the one now at last completed.

WALTER GELLHORN

June 1950

Index